Just-in-Time Manufacture

International Trends in Manufacturing Technology

Technology

JUST-IN-TIME MANUFACTURE

Edited by
Prof. C.A. Voss

IFS (Publications) Ltd, UK

Springer-Verlag
Berlin Heidelberg New York
London Paris Tokyo

C.A. Voss
School of Industrial and Business Studies
University of Warwick
Coventry CV4 7AL
England

British Library Cataloguing in Publication Data

Just-in-time manufacture.—(International trends in manufacturing technology).
    1. Factory management
    I. Voss, Christopher   II. Series
    658.5   TS155

ISBN 0-948507-49-7 IFS (Publications) Ltd
ISBN 3-540-17433-8 Springer-Verlag Berlin
ISBN 0-387-17433-8 Springer-Verlag New York

Phototypeset by Wagstaffs Typeshuttle, Henlow, Bedfordshire
Printed and bound by Short Run Press Ltd, Exeter

International Trends in Manufacturing Technology

The advent of microprocessor controls and robotics is rapidly changing the face of manufacturing throughout the world. Large and small companies alike are adopting these new methods to improve the efficiency of their operations. Researchers are constantly probing to provide even more advanced technologies suitable for application to manufacturing. In response to these advances IFS (Publications) Ltd is publishing a series of books on topics that highlight the developments taking place in manufacturing technology. The series aims to be informative and educational.

Books already published in the series are:

Robot Vision, Programmable Assembly, Robot Safety, Robotic Assembly, Flexible Manufacturing Systems, Robot Sensors (two volumes), Education and Training in Robotics, Robot Grippers, Human Factors, Simulation in Manufacturing, Automated Guided Vehicle Systems, and Robotic Welding.

Forthcoming titles include:

Artificial Intelligence in Manufacturing, and Programming Languages.

The series is intended for manufacturing managers, production engineers and those working on research into advanced manufacturing methods. Each book is published in hard cover and is edited by a specialist in the particular field.

This, the thirteenth in the series – Just-in-Time Manufacture – is under the editorship of Prof. C.A. Voss of the University of Warwick. The series editors are: Mike Innes and Brian Rooks.

Finally, the Publisher's gratitude is expressed to the authors whose works appear in this book.

Acknowledgements

IFS (Publications) Ltd wishes to express its acknowledgement and appreciation to the following publishers/organisations for granting permission to use some of the papers reprinted within this book, and for their cooperation and assistance throughout the production process.

Assocation for Manufacturing Excellence Inc.
380 W. Palatine Road
Wheeling, IL 60090
USA

American Institute of Industrial Engineers
25 Technology Park/Atlanta
Norcross, GA 30092
USA

American Production & Inventory Control Society
500 West Annandale Road
Falls Church,
VA 22046-4274
USA

American Society for Quality Control
230 West Wells Street
Milwaukee, WI 53203
USA

Automotive Industry Action Group
North Park Plaza
Suite 830
17117 W. Nine Mile Road
Southfield, MI 48075
USA

British Cast Iron Research Association
Alvechurch
Birmingham B48 7QB
England

IFS (Conferences) Ltd
35–39 High Street
Kempston
Bedford MK42 7BT
England

Institute of Management Sciences
290 Westminster Street
Providence, RI 02903
USA

Institute of Management Services
1 Cecil Court
London Road
Enfield
Middlesex EN2 6DD
England

MCB University Press Ltd
62 Toller Lane
Bradford
West Yorkshire BD8 9BY
England

Society for Advancement of
Management
2331 Victory Parkway
Cincinnati, OH 45206
USA

University of Nottingham
Department of Production
  Engineering
Nottingham NG7 2RD
England

Wright Publishing Co. Inc.
1422 West Peachtree Street
Suite 303
Atlanta, GA 30309
USA

CONTENTS

1. JIT Overviews

2. Manufacturing Techniques

3. Production/Material Control

4. Organisation for Change

5. JIT Purchasing

6. JIT in Action: Case Studies

Preface

Just-in-time (JIT) is a disciplined approach to improving overall productivity and eliminating waste. It provides for the cost-effective production and delivery of only the necessary quantity of parts at the right quality, at the right time and place, while using a minimum amount of facilities, equipment, materials and human resources. JIT is dependent on the balance between the supplier's flexibility and the user's flexibility. It is accomplished through the application of elements which require total employee involvement and teamwork. A key philosophy of JIT is simplification.

An ever-increasing number of companies are introducing JIT, with many more wishing to do so. This book has been designed for both groups of companies. It looks not only at JIT as a whole, but the various techniques of JIT. It also examines the implementation of JIT and includes case studies drawn from Japan, Europe, USA and others.

The company wishing to introduce JIT needs to know the answers to the following questions:

- What is JIT?
- What are the benefits of JIT?
- What are the techniques of JIT?
- How to implement JIT?
- What case examples are there of successful introduction of JIT?

This book has brought together the best writings on JIT in order to answer the above questions.

What is JIT?

JIT has its origins in the Toyota company in Japan. In the 1960s, Toyota worked hard on developing a whole range of new approaches to managing manufacture. The development of these approaches was hastened by the 'oil shock' of the 1960s. By 1972 these new approaches had begun to attract wide attention in Japan and in the mid 1970s other Japanese companies began to experiment with and adopt these approaches. At this stage and for some time later these were not known as JIT, but the approach was called 'Toyota Manufacturing System'. By the end of the 1970s the Toyota Manufacturing System had begun to attract attention in the West. One of the many elements of this system was a pull scheduling technique using 'kanbans' (Japanese for cards). The system first became known in the West as 'The Kanban System'. This was rather misleading as Kanban is only a small part of the total system, and very difficult to operate independently of a large set of other activities.

In the last few years, the approach has become widely known, particularly through Richard Schonberger's book 'Japanese Manufacturing Techniques'. As the approach has been adopted and adapted by Western companies, it has been given many names including 'Zero Inventories', 'World Class Manufacturing, and 'Continuous Flow Manufacturing'. (The latter term, used by IBM, is a more accurate description than the others.) However, the term that has now become most widely used to describe this approach to manufacturing is Just-in-time management.

Just-in-time management is not one technique or even a set of techniques for manufacturing, but is an overall approach or philosophy which embraces both old and new techniques. A full definition of JIT opens this preface. A more simple definition of JIT is:

An approach that ensures that the right quantities are purchased and made at the right time and quality, and that there is no waste.

Benefits of JIT

JIT is characterised by its ability to realise a wide range of benefits, often very rapidly. The short-term pay-offs often far exceed those realised by investments in sophisticated manufacturing technology. (JIT should not be considered a substitute for new manufacturing technology, but in many cases it is a very effective precursor.) In a recent survey of UK companies that have adopted JIT, the main benefits that had been achieved (in rank order) are shown in Fig. 1.

This book contains a number of case descriptions of successful implementation of JIT. Some examples of the benefits achieved are shown in Fig. 2.

```
1  WIP reduction
2  Increased flexibility
3  Raw materials/parts reduction
4  Increased quality
5  Increased productivity
6  Reduced space requirements
7  Lower overheads
```

Fig. 1 Ranking of the major benefits achievable using JIT

```
Inventory reduction
    WIP reduction of 84% in 4 weeks            Lucas
    $2 million reduction in inventory          Hewlett-Packard
    20% reduction in inventory                 Repco
    WIP reduction of 99%                       Lucas

Set-up time reduction
    60 minutes to 5 minutes                    Sumitomo
    30 minutes to 5 minutes                    Lucas
    Batch size of one                          Lucas

Space reduction
    Space reduction 46%                        Hewlett-Packard
    Space reduction 30%                        Rank Xerox

Productivity
    Increase 35%                               Lucas
    Reduction in direct labour 33%             Ford

Quality
    Solder rejects from 5000ppm to 20ppm       Hewlett-Packard

Investment
    Reduction in number of fork-lift trucks 52%   Rank Xerox
    Reduction in number of M/C tools 25%          John Deere

Cycle/lead times
    Lead-time reductions, 5 days to 5 hours    Lucas
    84% reduction in lead time                 John Deere

Other benefits
    Increased flexibility                      Lucas
    Build credibility 85% to 97.5%             Navistar International
```

Fig. 2 Typical benefits resulting from the successful implementation of JIT

Techniques of JIT

There are a very large number of techniques and approaches associated with JIT. In this book, examples are given of the major techniques. These fall into three main areas, as summarised in Fig. 3.

```
Manufacturing techniques              Inter-company JIT
    Cellular manufacturing                JIT purchasing
    Set-up time reduction                     Single sourcing
    Pull scheduling (Kanban)                  Supplier quality certification
      Group technology                        Point-of-use delivery
      Smallest machine                        Family of parts sourcing
    Fool-proofing (Pokayoke)
    Line stopping (Jidoka)            Organising for change
    U-shaped lines                        Quality
    Housekeeping methods (Seiri, Seiton)  Continuous improvement
                                          Enforced problem solving
Production/material control               Implementation
JIT – MRP
    Backflushing
    Flat bills of material

OPT
Schedule balance and smoothing (Heijunka)
    Under capacity scheduling
    Visible control
```

Fig. 3 Techniques of JIT

Manufacturing techniques

These are the techniques with which most companies will probably begin their JIT effort. JIT focuses on flow through the operation and cellular manufacturing is one of the core techniques that leads to flow. A key analytical technique leading to cellular manufacturing is group technology. These are supported by set-up time reduction, a key technique that enables manufacture to be in very small batch sizes. Small batch sizes in turn make continuous flow easier to attain. Once these have been implemented, the next technique that can be used is pull scheduling. This is normally known as Kanban, named after the Japanese for the cards that are often used in this system. There are two forms of Kanban, one-card and two-card. These core techniques are supported by a host of other non-core JIT techniques. These do not justify separate chapters in the book, but readers of the chapters and cases will find many of these mentioned repeatedly. They include the use of the smallest possible machine, the principles of 'Seiri' (putting away everything that is not needed), 'Seiton' (arranging those things in the best possible way), preventive maintenance, 'Pokayoke' (foolproof devices to prevent mistakes and defects), 'Nagara' (self-developed machines), 'Jidoka' (automatic stopping of the production equipment, automatically or by the worker, when abnormal conditions are sensed or occur), U-shaped lines (within a cell), and standardised packaging and containers.

Production materials control

Manufacturing planning and control systems require adaptation to a JIT environment. Materials requirements planning (MRP) and JIT can mutually support each other, but major adaptations must be made. These adaptations include 'backflushing' of inventory and flat bills of material.

Not all JIT techniques are Japanese. One particular technique that originated in Israel is 'Optimised Production Technology' (OPT). This is an alternative approach to scheduling to give JIT production in an environment characterised by complexity and known bottlenecks. JIT is greatly helped by reduction of variability, and one of the core JIT techniques is the smoothing of production schedules. To support schedule smoothing, techniques of balancing production ('Heijunka'), have been refined. As with manufacturing techniques, there are a number of other techniques which do not deserve a chapter of their own but are worthy of mention. These include under-capacity scheduling (deliberately scheduling slack time during each shift so as to allow for problem rectification), and visible production control (minimising control paperwork).

JIT purchasing

JIT principles can be applied on an inter-company basis as well as inside a company. JIT purchasing requires that goods are supplied in small quantities, in exact amounts, at frequent intervals and at 100% quality. There is a wide range of specific techniques related to JIT purchasing. These include single sourcing, use of standardised containers, supplier certification, point-of-use delivery, supplier controlled pick-up, purchasing Kanban, early supplier involvement in product development, linked information systems, family of parts sourcing, and above all mutual trust.

Implementing JIT

The prime objective of JIT is the elimination of waste. Implementation focuses on making this happen. To implement JIT effectively three areas must be considered. The first is that of quality. Very high levels of quality are a necessary condition for JIT. Any JIT effort must have an accompanying quality improvement programme. One core JIT philosophy is that of continual improvement ('Kaizen'); extra improvement should always be sought. There is no best, only better. This goes hand-in-hand with the technique of enforced problem solving.

A commonly used analogy is that of a ship sailing on a lake (see Fig. 4). The rocks represent the problems that underlie any manufacturing environment. By deliberate removal of slack (e.g. buffer stocks), we expose the rocks. The JIT philosophy is that we must solve those

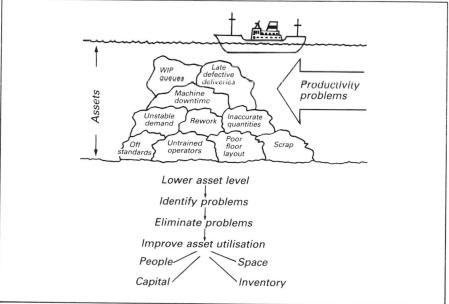

Fig. 4 Enforced problem solving

problems immediately and solve them permanently. When this is done, we can remove a little more slack and expose some more problems. Implementation of JIT is also concerned with the most important asset – people. JIT implementation programmes pay particular attention to the involvement of people through such techniques as pushing decision making back down to the cell level, use of multi-functional workforce and small group improvement activities.

Case examples

One of the most effective ways of learning about new approaches and techniques is to study successful applications. Included in this book are a number of case studies and other examples drawn from a range of industries and countries. JIT has been most widely applied in the automotive industry and the electronics industry. The cases reflect this in that the best examples will be in industries with the greatest experience. JIT is not confined to these industries. For example, many process industries are adopting selected techniques, in particular set-up time reduction. JIT has now spread from Japan to many countries. This is reflected in this book, where cases and illustrations are drawn not only from Japan but from Australia, West Germany, the UK, the Netherlands, South Africa and the USA.

Summary

JIT is both a philosophy that can help companies change the way that they manage manufacturing and to become more competitive. This book is designed to answer the many questions companies may have, such as what is JIT, what are the benefits, what are the techniques, how to implement JIT and what is good practice? It is hoped that this book will contribute to the understanding and the increased adoption of one of the most effective and exciting approaches available to manufacturing companies.

Professor C.A. Voss
University of Warwick

1

JIT OVERVIEWS

JIT is an approach and a philosophy which covers a wide range of techniques and considerations. This section offers the reader an overview of JIT against which can be set the detailed considerations in subsequent sections of the book. It emphasises the role of JIT in the elimination of waste and the need to replace complexity with simplicity. It is intended to provide an introduction to the various techniques of JIT and to demonstrate how they fit together.

JUST-IN-TIME
REPLACING COMPLEXITY WITH SIMPLICITY

R.J. Schonberger
Schonberger and Associates Inc., USA

JIT, TQC and TPM are replacing the over complex methods of the past in which a high level of waste is permissible and, indeed, the 'norm'. The accent now is on flexibility, and process layouts are being replaced by cells and flow lines manufacturing production families. Shorter production runs and fast changeovers are also required, with maintenance now being more important than ever before. Workstations are moved closer together in distance and time to cut down flow distance and to involve operators in problem solving. In a new approach, manufacturability and quality are designed into the product. Expensive buffer stocks are replaced by frequent, small deliveries of supplier-certified quality parts. Kanban or, better still, visual signal systems are used to 'pull' parts only as required. Contrary to popular belief, cost, quality, time and flexibility are not in opposition. In JIT production, the causes of stoppages and slowdowns are recorded and the data so generated should be put to use in order to get the best from the system. By using JIT, a few companies have achieved a five- to ten-fold reduction in lead times. In the JIT environment, computer support is not as vital as in job-lot manufacturing, where production and inventory control is crucial to high performance. Indeed, one can envisage computers being used for major event planning only every few weeks.

Just-in-time (JIT) production and its companions, total quality control (TQC) and total preventive maintenance (TPM), are brashly elbowing their way into manufacturing plants throughout the world – galloping over the carefully fashioned manufacturing management systems that had been in place. The old systems now appear to be overly complex and to contain seeds of deterioration: built-in high levels of permissible waste.

The waste is found in our conventional approaches to quality, design, purchasing, job assignments, plant configuration, equipment selection, maintenance, purchasing, scheduling, accounting, product-line development, material handling, material control and shop-floor control. The waste is found – and the interest in JIT is appearing – in nearly all manufacturing industries throughout the world.

There are still misunderstandings of JIT's objectives. For one thing, there is too much spending on equipment without regard for the flexibility of the plant. There also are some fundamental questions and issues to be resolved, such as whether to throw out or patch up state-of-the-art computer-based systems of inventory and shop-floor control (such as MRP).

While the issues have not been resolved, there is a persuasive argument that existing subroutines of MRP may be harnessed as an effective approach to *major event planning* (MEP).

JIT implementation

Just-in-time projects are being launched in most of the better known industrial firms in the world. The projects turn into JIT campaigns, and the campaigns link with total quality control to become, as many are saying, the 'new way of life' in manufacturing.

In a few companies, results are in the order of five- to ten-fold reductions in lead time (throughput time). For example, several of Hewlett-Packard's and Omark Industries' plants have achieved these levels of improvement.

Just-in-time was firmly in place in numerous Japanese plants 18 years ago, but its beginnings elsewhere date back only to about 1980. General Electric perhaps was the first, outside of Japan, to mount a JIT campaign.

Two GE plants had JIT projects in 1980; there were 10 GE projects (or plants with projects) in 1981, 20 in 1982, and 40 in 1983. GE's JIT efforts are in a variety of types of manufacturing: locomotives, dishwashers, huge vacuum circuit breakers, jet engines, switches, generators, lamps (light bulbs), coffee makers, and more.

These products (and their components) are high and low volume, high and low variety, discrete and process, make-to-order and make-to-stock, smokestack and light assembly.

Shop-floor organisation

JIT is usually most successful when the focus is on moving workstations close together. This act reduces several forms of waste and cost. These include:

- Transit time.
- Queue time.
- Transit inventory.

- Queuing inventory. This can be avoided by moving workcentres close together so that it becomes natural to try to synchronise their schedules, overlap work orders and wipe out queuing inventories.
- Space.
- Time between occurrence and discovery of defects.
- Poor environment for problem solving. A corrective action here is to get workstations linked together in time and distance so that operators will have reason to embroil themselves in problem solving.
- Poor environment for shop-floor communication and teamwork. By removing the walls of inventory, people can see each other, share successes and failures, come to each other's aid and learn each other's jobs.

There are many JIT companies that recognise the power of moving workstations together. At one GE plant the JIT effort features 'moving a machine a day'.

At Omark's Guelph, Ontario plant, which makes saw chains, the flow distance was 2620ft in November 1982, and flow time was 21 days. Metal forming machines have since been moved together, squeezing out work-in-process inventory and reducing the distance to 173ft and the time to three days. When the time drops to one day, Omark's new marketing strategy – to fill orders from the factory instead of from finished goods warehouses – will become viable.

In companies that fill orders from warehouse stocks, backorders are common because the warehouse, while having a *lot* of stock, often does not have the *right* items. Omark may actually be able to improve upon its past customer service *without carrying finished goods inventories.*

In Buick's Flint, Michigan, manufacturing centre, the 'Buick City' project features relocating virtually every machine and workcentre and replacing rail sidings with truck doors. The plan is to get frequent 'just-in-time' raw and component parts deliveries from suppliers and fashion those materials into cars quickly by flowing them through closely coupled machine centres.

Rockwell International, Telecommunications Division, operates a sizeable plant with standard metal forming equipment near Dallas. It's a job shop: many part numbers, with irregular orders for each one. For over two years Rockwell has been tearing apart its machine centres – degrease, silver solder, sheet metal, spot weld, shear, drill, grind, NC lathe, and so forth. The thousands of part numbers were organised into production families by sorting route sheets (a manual effort since part numbers were not coded to permit computer sorting).

The next step was to begin moving machines out of the traditional process layouts and into cells. A cell for making waveguide parts came first, and now several cells are operational, each resulting in greatly reduced throughput time, idle inventory and rework/scrap costs. Called cellular manufacturing, this concept is the basis for Just-in-time production where there are many part numbers and small production runs.

Small job shops use something resembling cellular manufacturing without realising it, simply because they have never become so large as to have multiple machines of the same general type. The temptation to organise machines into shops – the process layout – is fortunately avoided.

Thus, when small companies implement JIT, relocating equipment is not normally going to gain much. Instead, the small job shop needs to build on its flexibility. This can be done in several ways, which include:

- Reducing set-up times.
- Becoming fanatical about machine maintenance.
- Employing overlapped production, in which a job is split so that it is in process in several different stages of production at the same time.
- Perfecting a system to guarantee quality to its customers.
- Chopping lead times by aggressively seeking long-term contracts, with customer requirements built into the job-shop schedule.
- Developing a competitive analysis capability as a cornerstone of product and process development.

Preparation for automation

JIT has other names, such as 'zero inventories', 'material as needed', 'stockless production' and 'continuous flow manufacturing' (CFM). CFM, perhaps coined at IBM, conjures up a vision of products not stopping. Does this mean a factory filled with conveyors and automated material handling?

It should not, because automated handling and automated processing cost money and may make the system less flexible. If flexibility does not increase, then something is wrong with the JIT campaign.

Machines (including robots) are several orders of magnitude less flexible than people are. Just-in-time and automation strategies are eminently compatible, but only if the automation is introduced as solutions to problems.

This usually means piecemeal automation and robotics, which is *problem-pull* rather than *technology-push* automation. Furthermore, in the problem-pull mode much of the automation may be in the form of small homemade devices to measure, hold, identify, locate, index, receive, align, orient, pull forward, and so forth.

Automatic inspection, and perhaps servo mechanisms (to automatically adjust machine settings), are also likely. The purpose of these devices is to make it easier for people to work safely and efficiently in a uniform cycle time, without error. They also pave the way for further automation and perhaps robots.

It's often not difficult to justify robots as replacements for high-wage humans. But how many companies, in their economic analyses, crank in such intangibles as the robots' lack of capacity to think, to diagnose, to solve problems, to innovate and to self-adjust?

We hear a good deal about big-splash automation projects, such as IBM's PC plant in Boca Raton, Florida, and Apple's Macintosh plant in Fremont, California. These plants appear to be enormous financial successes in that the investment in equipment is paid off many times over. It is easy, however, to overlook the uniformity and simplicity of the products themselves.

The marketing concept is to offer standard 'vanilla' products, with no features. The developmental engineering concept is supportive; IBM, Apple, and most of our other leading electronic 'box' manufacturers have learned how to design products with virtually no screws. Manufacturing is being reduced to push and snap, which makes it easy for a robot to make the product – or for hard automation or humans to make it, for that matter.

The better companies are learning to design products for quality and manufacturability and to push quality assurance back to the suppliers. The threshold of high performance is rising accordingly. IBM, Apple, and other leaders no doubt will also learn how to employ the problem-push automation approach, which favours incremental (perhaps rapid) upgrading of processes, instead of large-leap process introductions followed by periods of complacency.

The thoughtfully planned factory of the *near future* should preserve and steadily enhance flexibility through a combination of: simple, low-cost, dedicated machines and devices; some use of costly equipment – it's often unavoidable – but equipped to move easily and to be set up quickly; and humans.

Changes in quantities or model mix should be taken in stride. Even changes from one generation of products to the next should be handled without major upheaval: discard the cheap machines, reprogram the costly machines, and put the adaptable human labour on the tasks of diagnosing and solving problems anew.

Supplier quality assurance

Just-in-time purchasing seems to be the aspect of JIT that gets the most press attention. The attention focuses on efforts to get frequent deliveries in small amounts from nearby suppliers.

More important than the issue of delivery frequency is that of supplier-assured quality. Supplier deliveries must be of good quality and on time, or the customer's plant will be plagued by severe stoppages, which cannot be tolerated if the customer plant has squeezed out most of its protective buffer stocks.

Most of the top companies *are* working on supplier development and supplier-certified quality, but the old adversary relationship between customer and supplier will take years to fix. Reductions in purchased material stocks are coming slowly, as one supplier at a time is covered by a long-term contract and supplier process controls, instead of sporadic purchase orders and bids and receiving inspections.

The quality movement began before the JIT movement – in Western industry as well as in Japan. Our quality and JIT movements need to merge and become indistinguishable in the West as they are in Japan. This is beginning to occur in some companies like IBM, GE, Hewlett-Packard and the auto companies. A vigorous quality effort accelerates JIT implementation, and vice versa.

On the one hand, the best medicine for lowering the cost quality is cutting inventories and thereby the number of units to be reworked or scrapped when a lot is bad. On the other hand, the best grease for JIT implementation is defect-free materials so that extras are not needed.

A basic element of JIT is recording all of the causes of production stoppages and slowdowns (and removing waste so that stoppages and slowdowns *do* occur regularly but in a controlled fashion). Western JIT plants often are not taking advantage of this data-generation opportunity, perhaps because it means giving shop-floor operators some new responsibilities.

JIT lights

Having operators record quality levels on process control charts is often the first move in the right direction. The next step might be to add trouble lights activated by operators. Typically, there is a red light to turn on to trigger and signal a line stoppage for bad quality, machine trouble, lack of parts, etc., and a yellow light to turn on when there is a slowdown but no need to stop the whole line.

These trouble-light systems are found widely in Japanese industry, having spread from the auto industry to many others in recent years. Some of our Western factories have had lights for years, unrelated to Just-in-time.

For example, in tyre plants there sometimes are lights on the press-cure machines, and in printed circuit board shops there sometimes are lights on final board test equipment.

JIT/TQC lights are different. They *may* summon help; they *require* that a record be made of the reason why. The different types of problems are tallied, often on a blackboard or clipboard.

The group leader or supervisor uses the data to prioritise problems and to assign project teams to solve the problems. If they exist, quality control circles can perform some of the data analysis, diagnosis, prioritisation, and problem-solving.

So far, there may be 50–100 plants in North America that have installed warning/stop light systems as part of a JIT campaign. Many have not added the tally boards and therefore are missing out on the most valuable benefits. One well publicised case in which the lights are used correctly is at a Hewlett-Packard plant in Greeley, Colorado.

Kanban

Kanban, a Japanese word meaning card, is the user's signal of a need for another container of parts from the source of the parts. Kanban 'pulls' the container; JIT prohibits the source of the parts from 'pushing' them forward, only to sit idle at the user's workstation. Not many companies in the world use Toyota's well publicised dual-card Kanban system. Renault in France is one company that does.

Single-card Kanban, or better yet, visual signal systems (both also perfected and widely used by the Toyota family of companies), are more common. The author estimates (based on extrapolation, not an actual count) that over 100 Western JIT companies now use such signal systems. For example:

- Within GE's lamp division, consisting of multiple plants in the State of Ohio, truck drivers from component plants collect kanban and empty containers when they unload. The cards signal which components are to be delivered next trip.
- At Hewlett-Packard, Fort Collins, the signal for a subassembly shop (making computer system modules) to send another plastic tub of parts forward is removal of the present plastic tub of parts from a sensing platform.
- At Hewlett-Packard, Greeley, an empty 'kanban square' outlined in yellow tape is the visual signal for the preceding workstation to forward another disk drive unit.
- At one semiconductor plant in France, the JIT group cut throughput time and inventories of wafers in process in half by employing a more generalised signal system: a daily meeting of all the fabrication centre supervisors to make sure that lot movements are synchronised; no more making and pushing lots forward unless the next stage is ready to process them.

When production is conveyor-paced, Kanban is present in disguise. The chain pulls the workpieces forward, one unit at a time. Conveyorised production lines have been used in Western industry for years.

Now some of these lines are being outfitted with trouble and stop lights and with charts to record hourly production and problems. The next steps are to extend the assembly-line ideals back into prior stages of manufacture:

- Cut from large-lot to small-lot or to one-piece-at-a-time production lines in subassembly and even fabrication.
- Synchronise subassembly, fabrication and supplier schedules, expressed as daily rates, with final assembly rates.

Many companies have made excellent progress in cutting lot sizes; synchronising schedules is coming slowly.

Production runs

Toyota and the Japanese auto industry developed JIT, and then the
concepts spread to most other Japanese industries. This is the pattern
outside of Japan as well. We hear a good deal about JIT in the auto
industry, but not much about existing vigorous JIT campaigns
elsewhere.

There are enlightened companies in weaving, tyres, pharmaceuticals, sandpaper, paper and packaging materials, bottling, wire,
semiconductors, canning and plastics, to name a few. In these
industries products tend to be made in volume and to follow standard
flow paths, which are some of the characteristics of the process
industry. But different sizes, colours, grades and formulations must be
produced, and generally large inventories are built up between changes
and between production stages.

In pursuing JIT these producers find that they must move toward
shorter production runs (the opposite of conventional thinking) and
fast changeovers. Tight process control, total preventive maintenance,
frequent small deliveries of raw materials and synchronised scheduling
are required to keep the plant running as lead times and buffer stocks
are cut.

In many cases JIT begs for smaller, more flexible equipment in
multiple copies so that there can be provider machines right next to user
machines throughout the plant. For example, most Western tyre plants
cluster first-stage tyre-building machines together in one shop,
second-stage machines in another shop, and so forth. Just-in-time
requires moving these machines into flow lines.

The immediate results will be sharp cuts in numbers of tyres between
stages. This should yield competitive advantages. Deep reductions in
cost of quality come right away: few bad tyres and blems to inspect and
rework when final inspections reveal problems.

In semiconductor manufacturing the situation is similar. Though the
equipment is far more complex, immobile and costly to move into flow
lines. It appears that not even the Japanese semiconductor manufacturers have had much success in setting up flow-line production in their
wafer fabrication plants.

The door is open for somebody to figure out how to do so; a few flow
line modules of the most mobile and flexible machines would be the
place to start. Ultimately, maybe the immobile machines need to be
made smaller and cheaper so that dedicated lines are affordable for
'class A' part numbers; also there is a need for equipment configured
for quick utility connects/disconnects.

Firestone, Frito-Lay, Intel, Burlington, RJR Archer (Reynolds
Tobacco), McNeil Pharmaceuticals, Procter and Gamble, International Paper, Raychem and 3M are a few of the semi-process-oriented
companies that have launched JIT campaigns. At the other end of the
spectrum are low-volume, high-variety manufacturers. Most are very

small: machine shops, foundries and other general-purpose job shops. These kinds of firms constitute the bulk of industrial organisations, and so far the small ones are mostly not on the JIT roller coaster. Their customers are slowly getting them involved; the starting point usually is placing the responsibility for quality assurance on the supplier.

The situation is different for the large job shops, which often make end products like locomotives, aircraft, ships and large computers. A few aerospace companies (for example, Hughes) are quite involved in JIT. A contract may call for only one radar system or airplane per week or per month, but some at Hughes see even that as offering possibilities for regularity and synchronisation in scheduling the thousands of production steps.

In shipbuilding a revolution of sorts (begun in Japan) has been in progress. Some of its features are synchronised scheduling of standard work packages, standard flow paths, and cellular grouping of processes of machines.

These techniques have helped some shipbuilders (e.g. Todd Shipyards) to reduce the time to build ships from months to weeks. This is JIT implementation in the most complex of production and inventory control environments.

Finally, there are the mass producers of industrial and consumer tools, appliances, furnishings and so forth: cameras, cassette recorders, toys, shoes, luggage, washing machines, generators, motors, desks, power supplies and thousands more.

This manufacturing sector seems to absorb JIT most easily, because much of the production is labour intensive. Since labour is flexible, the cost of JIT in these industries may be modest: mainly education, training, and coordination. In advanced cases, employing automation, robotics, flexible manufacturing systems, and computer-aided engineering and processing, the costs are much higher.

With the automation and computer-integrated manufacturing wars on the horizon, the winners will be those that preserve flexibility so that features, models and even whole product lines may be changed at low cost. If this sounds like an impossible challenge, we should realise that companies like Hitachi, Canon, Bridgestone and Honda already know how to cut costs, improve quality, shorten lead times and quickly evolve to new products.

An unfortunate myth, promulgated in textbooks, has it that these competencies – cost, quality, time and flexibility – are in opposition.

Role of computers

The simplicity and common sense of the JIT concepts make them easy to accept and to implement. While we proceed to install simplicity, what do we do with the complexity? Much of it – complicated stock control systems, automated stockrooms holding work-in-process (WIP) inventories, elaborate scheduling and order tracking systems – tends to melt away as JIT techniques are implemented. The computer systems that manage the complexity are in the spotlight. What should we do with them?

To answer this, we should understand that the Japanese (and the Europeans) are many years behind the North Americans in using computers in business and industry. Leading US manufacturers have been perfecting material requirements planning (MRP) for 20 years or more, while their Japanese competitors have mostly ignored it.

Now many Japanese companies have MRP projects in progress and in some cases up and running. While we tend to think of MRP first and foremost as a system for planning and controlling orders on the shop-floor and in the supplier system, MRP also performs certain medium-range planning tasks with great efficiency. These medium-range planning activities may be referred to as major event planning or MEP.

Major events in a manufacturing plant include new products, new model mix, new schedule, major engineering changes and new processes. The introduction of any such major events has spillover effects in all directions: correct timing of the necessary tooling, equipment, space, labour and labour assignments, made and bought parts, even cash flows.

MRP in Japan

In the past the Japanese have made the necessary calculations via primitive manual methods; more recently they have used computers, but usually independent programs and no common database. The Japanese have a good deal to learn from the Americans in how to make the calculations efficiently, working from a common database: a master schedule file, bill of materials file, and other resource files in the computer. Western MRP and MRP II (manufacturing resource planning) systems provide this capability.

While the Japanese are implementing the MEP subroutines, the North Americans should be discontinuing some of the uses of the bread-and-butter MRP subroutines for shop-floor and vendor scheduling of orders, order control, inventory control, labour reporting and material reporting. This is reasonable for any product that is put onto a daily rate instead of separate work orders and purchase orders.

In the low-volume non-repetitive ends of a business, the need for MRP controls probably will persist. However, the amount of processing should be greatly diminished as our fabrication equipment and subassembly workcentres are reconfigured for cellular manufacturing and dedicated flow lines. There will no longer be a need to schedule and track inventory movements from machine to machine and station to station; only to schedule the cell or line.

In some cases kanban signals from one cell to another can even eliminate the need for independently scheduling the cells and dedicated flow lines. Of course, as we move into the automated robotised factories of the distant future (distant because of the expense), the computer takes on new roles.

The new roles will not centre around traditional notions of manufacturing information systems as much as direct process control. Machines talking to each other need not talk in human-readable languages.

Concluding remarks

It was quite proper to heavily emphasise computer support in the past, because production and inventory control was vital to high performance in a job-lot manufacturing environment.

It is not so vital in the emerging JIT environment, and there is the risk that early plans for computer support will need to be undone later.

In the well-managed factory of tomorrow – today in a few cases – major event planning on the computer will occur only once every few weeks. Much of what happens on the shop-floor will be managed 'by sight' (a GE term) and by those closest to the action – the operators and their supervisors.

JAPANESE JIT MANUFACTURING MANAGEMENT PRACTICES IN THE UK

C.A. Voss
University of Warwick, UK

Japan is perceived in most advanced countries as the world leader in
the introduction and exploitation of new management practices,
particularly in the area of manufacturing. In an attempt to match
Japan's economic progress other countries are seeking to introduce
many of the systems which are believed to be the basis of Japanese
success. In this paper, certain aspects of a UK manufacturing
company, which has adopted a number of practices following studies
undertaken in Japan, are compared with a Japanese owned company
located in the UK.

The first report of Japanese manufacturing management practices (as
opposed to management practices in general) to make a major impact
on the West was probably that by Sugimori et al.[1] in 1977. They
described various approaches at Toyota including the Kanban system,
reduction in set-up times and 'Just-in-time' production. They also
hinted at the impact of quality circles.

Since then there have been a stream of reports and publications on
Japanese manufacturing management. These in general can be divided
into quality and quality circles[1-4], inventory reduction and the Kanban
system[1,4,5], labour productivity, Japanese culture and management
style; for example, 'Theory Z'[6-8], investment, robots and
automation[4,9-12] and general manufacturing management[10,13].

In the wake of this upsurge of interest in Japan, and the increasing
competitiveness of Japanese manufactured goods, many organisations
have examined these practices with a view to transferring them to their
own country. The technique of quality circles is probably the practice
that has transferred most widely and with least adaptation[2,3]. A
second course of transfer has been the establishment of Japanese
owned factories overseas. In these there has been a much more
complete transfer of management style, maintained by the presence of
Japanese managers[4].

The study of the transfer and applicability of Japanese manufacturing practices has in general focused on the general, managerial approach rather than specific manufacturing/operations management practices[6,14]. The one exception to this is the area of quality circles. The objective of this study has been to examine the spread of specific Japanese manufacturing practices to the UK. Two means of transfer have been considered. The establishment of Japanese owned and managed factories and the attempted adoption of Japanese manufacturing management practices in the UK. This paper reports a small study in which one Japanese owned and one UK manufacturing company were examined.

JIT management practices

Popular opinion often views the important Japanese manufacturing practices as being the use of robot and quality circles. Hayes[10] reported that these practices were used but that:

> "For the most part, Japanese factories are not filled with highly sophisticated equipment . . . the general level of technological sophistication that I observed was not superior to (and generally lower than) that found in US plants."

> ". . . the famed 'quality circles' did not appear as influential as I expected . . . most of the companies . . . had enviable reputations for high quality products by the time they adopted QCs."

He did observe across a wide range of plants a pattern of common manufacturing management practices. These can be summarised as:

- Clean orderly work place.
- Minimised inventory.
- Problem prevention.
- 'Pursuing the last grain of rice'.
- Thinking quality in.
- Equipment policies.

In addition to these practices, including quality circles and automation (robots), various authors have identified a number of different practices including the Kanban system, MRP (a Western practice), worker ability to stop the production line, managerial attention to detail and 'treating operations as stragegy'[15].

Methodology

For the pilot study two firms were chosen. The first is a Japanese manufacturing company operating a facility manufacturing consumer electronic products including television and tape recorders. The second is a European company manufacturing mass-produced consumer goods. The latter, which has a number of UK plants, had sent a number of groups of managers and engineers to Japan and was in the process of introducing some Japanese manufacturing practices. Both companies could be considered successful and both held substantial shares of the UK market.

A list was prepared of all the manufacturing practices reported by various authors. Each company was visited in 1982 to determine whether the practices were used at the factories of the company. For each practice the use or lack of use was observed and this was reported.

Table 1   Transfer of Japanese JIT manufacturing management practices to the UK

	Observations in UK	
Management practices	Japanese owned factory	UK factory trying to adopt Japanese practices
A. Practices reported by Hayes[10] to be prime factors of Japanese success		
1. Clean orderly workplace	Yes	Yes
2. Minimised inventory		
Little inventory on shop-floor	Yes	Yes
Goods delivered when required (just-in-time) in small lots	Yes	No
Rapid set-ups	Yes	No
No buffer stocks	Yes	Some reduction
Clearly marked inventory	Yes	Yes
3. Problem prevention		
Preventing machine overload	Not observed	Not observed
Monitoring systems	Not observed	Not observed
No crisis atmosphere		
Long stable master schedule	Yes	Yes
Capacity based on actual measures	Yes	Yes
No overload	Yes	Yes
Minimised changeovers	Yes	No
4. 'Pursuing the last grain of rice'		
All problems considered important	Yes	Not observed
Strive for continuous incremental improvement	Yes	Not observed
5. Thinking quality in		
Building quality into design	Done in Japan	No
Training of all workforce	Yes	No
Workforce do inspection (checking)	Yes	Some
Feedback to workforce	Yes	Little
6. Equipment		
Equipment independence	Yes	No
Well maintained standard machinery	Yes	Partly
Excellent material handling	Yes	Partly
B. Practices considered by Hayes not to be prime factors of Japanese success		
7. Quality circles	No	Yes
8. Robots	No	Yes
C. Practices mentioned by other authors		
9. Kanban	No	No
10. MRP	No	Yes
11. Worker ability to stop line	No	No
12. Attention to detail	Yes	Yes
13. 'Operations as strategy'	Not observed	Yes

Where it was not possible to determine whether the practice was used or not it was reported as 'not observed'. Four visits were made, including a full site visit, and there were discussions with senior and middle level manufacturing managers.

Results

The observations at the two companies ae summarised in Table 1. Some of the specific observations are described below.

Clean orderly workplace

Both companies had very tidy factories, though the Japanese factory was not as immaculate as a UK computer manufacturer recently visited by the author (this had wall-to-wall carpet!). The UK factory was very much tidier and cleaner than it had been five years previously.

Minimised inventory

The Japanese factory had exceptionally low levels of inventory. Just-in-time practices were applied to purchase goods. Very close relationships had been forged with suppliers, much material was locally sourced (70% by value was UK manufactured). For the main components, delivery was normally 24 hours before use. A maximum of four hours of stock was kept on the line. There were no in-process buffer stocks, though a ten day buffer of imported components was maintained because of uncertainties in customs clearance at the docks. The system was so finely tuned that heavy snow during the winter, which cut off supply lorries for 48 hours, caused the factory to halt. There was little machinery that required setting up; the exception being component insertion machines which were claimed to have very rapid set-up times (set-ups were not observed).

The UK company had successfully adopted some practices. The level of inventory on the shop-floor was impressively low and was very clearly marked and located. The company had begun to reduce its process and component buffer stocks, but had a considerable way to go. Despite management statements about intention to reduce set-ups, and hence batch sizes, no evidence was seen of this.

Problem prevention

Problem prevention policies, as described by Hayes, proved difficult to observe in practice. One area where observation could be made was policies concerning the Master Production Schedule (MPS). The Japanese factory endeavoured to maintain a frozen MPS over a three month horizon. This had recently been reduced because of major changes in the mix demanded in the marketplace. All jobs were measured on MTM standards and output was 98.5% of standard. There was no evidence of overloading. Changeovers were minimised with a targeted four-day run per model.

A very similar pattern was observed in the UK company except that a wide variety of models were produced simultaneously on the line with no obvious adverse consequences.

'Pursuing the last grain of rice'

Hayes used this expression to describe how Japanese manufacturing managers considered their responsibilities. This was broken down into two elements. First, managers (and workers) consider all problems to be important and, second, there is a continuous striving for incremental improvement. Interviews with the (British) manufacturing manager at the Japanese factory showed this approach very much present. He made a number of statements such as "all problems are the managers' problems". When asked how he spent his time, he indicated that much of his managerial effort was spent pursuing incremental improvements, particularly in manning levels on the line.

Such views and approaches were not openly expressed in the UK company, though it was not possible to determine whether they were in fact present but less visible.

Thinking quality in

The first impression on visiting the Japanese factory was of commitment to quality. The entrance lobby was full of posters painted by the workforce on the subject of quality, and quality was never out of the conversation. When asked about responsibility, every manager replied that quality 'was everyone's responsibility'. Each sub-group was totally responsible for the quality of the work that it produced. There was 100% testing at frequent intervals on the line. Testing was performed by one of the work group, not a quality inspector. (The quality department was concerned with design of tests, inspection of parts in goods inwards and final testing only.) Detailed attention was paid to quality in the design of the process. Each workstation had a small box for accidentally dropped parts; these parts were retested before being used. The overall quality targets were explicit and known by all the workforce. The quality performance was measured for each individual workgroup and worker. These were fed back to the workforce, who were rewarded on their individual quality perform-ance. Strong emphasis was placed on parts and material quality. Parts per million level were specified for components, and quality was placed above price in purchasing criteria. The Japanese company claimed that its field reject rate was only 15% of its UK competitors', though slightly higher than that of Japanese produced goods.

The UK company had always had a high standard of incoming parts and material inspection and had begun to adopt some of the Japanese practices. In some areas, the work of inspectors had been significantly reduced and the amount of worker inspection increased. There was some feedback to the workforce, but not to the extent of the Japanese company.

On the product design side, the Japanese products were clearly designed to maximise quality, though all product design originated in the Japanese parent. The UK company's products seemed designed to meet minimum quality specifications rather than to maximise quality.

Equipment policies

Hayes asserted that Japanese companies tended to try and maintain equipment independence by manufacturing much of their own process machinery. Some evidence of this was found in the Japanese factory where the component insertion machines had been manufactured by the Japanese parent company. Hayes also indicated that automation consisted of simple material handling equipment used in conjunction with standard processing equipment – the machinery itself being well maintained and smooth running.

The above is precisely the pattern that was observed in the Japanese factory. The material handling was highly automated and the most sophisticated equipment was automatic component insertion machines.

In contrast, the UK company purchased all its process machinery from outside suppliers. The material handling equipment when used was good, but was not as extensive as the Japanese factory – fork-lift trucks were much in evidence. Similarly, machinery seemed well maintained, but quite a number of breakdowns were observed during plant visits.

Robots

The Japanese factory did not have any robots or highly sophisticated automated machinery. It was a very labour-intensive manufacturing process. In contrast, the UK company was in the process of a major automation programme. A number of robots had been installed and more were on order.

Quality circles

The UK company had a mixed record with quality circles. On one site they had been successfully adopted; on another they had been tried and failed and on a third site they had not been tried. There were varied opinions for the failure at the second site; some blamed it on union resistance, others indicated that the company had tried to push them on the workforce rather than allow voluntary development.

The Japanese company did not use quality circles. When asked why, the management replied that they felt it was too early to introduce them; 'maybe in a couple of years' time'.

The Kanban system

The Kanban system is effectively a production control system that pulls jobs through a shop rather than pushes them in[5]. The Japanese factory did not use the system. The UK company had considered the system, and in one factory a group of pallets had the word kanban painted on them. However, there was no evidence of the use of the system.

Materials requirements planning (MRP)
MRP is not a Japanese-originated system (strictly speaking neither are quality circles), but some authors have reported it as being widely used in Japan[16]. The Japanese factory had a simple manual production planning and control system, which could not be considered MRP. The UK company used an MRP system. This system seemed to be well developed and working effectively.

Worker ability to stop line
Sugimori et al[1] reported that workers on the Toyota production line could stop the line if they had problems or failed to complete a task. This facility was not observed at either the Japanese factory or the UK company.

Attention to detail
Japanese manufacturing managers are reported to pay great attention to the detail of the manufacturing process[15]. Though difficult to observe in a short period of time, managers at both the Japanese and the UK companies gave the impression of being intimately involved with detail.

'Operations as strategy'
Wheelwright[15] asserted that Japanese companies could be differentiated from most US companies through "a profound commitment to what can only be called a 'strategic operations policy'." In the Japanese company it was not possible to observe this as such policies would be originated in the Japanese parent. In the UK company there was evidence of commitment to a strategic operations policy. This was confirmed through discussions with various managers who indicated that such policies existed.

Discussion
From Table 1, the first conclusion to be drawn is that the Japanese company had brought to the UK virtually all of the manufacturing practices reported by Hayes as being of prime importance, and in addition had neither brought those practices dismissed by Hayes nor those mentioned by other authors (except 'attention to detail'). In contrast, the UK company had adopted quality circles, robots and MRP; and had flirted with the Kanban system. In addition, the UK company had tried to adopt Japanese practices in most of the areas listed by Hayes, though usually not as well or as comprehensively as the Japanese company.

Why was the UK company trying to adopt quality circles and use robots when the Japanese company was not? A possible explanation may lie in the use of labour. The Japanese company started with a greenfield site and since its inception has maintained tight control of manning. The UK company, on the other hand, has long-established factories which had, by any standards, been heavily overmanned. It may well be easier to automate a process and discard virtually all the manpower involved through redundancy than to take the same overmanned process and reduce the manning to efficient levels without changing the process radically (even if the latter is more cost effective). Second, quality circles have as great an impact on labour motivation and productivity as on quality. It is possible that they were introduced for the former reasons rather than to improve quality.

Some of the areas warrant closer inspection. Many companies in the UK have examined Japanese inventory reduction techniques such as 'Just-in-time' and have rejected them as being inapplicable to the UK. The rejection has usually been on the grounds that British suppliers are unreliable and the labour force cannot be trusted.

The evidence from this study indicates that these are weak arguments. The Japanese company has managed to select good suppliers and to develop the necessary relationship (for example, through giving suppliers stable call-off schedules), so that just-in-time arrival of goods works. The Japanese factory was designed to have no buffer inventory. The UK company had begun questioning the assumptions behind its buffer inventory and was rapidly reducing stocks.

In 'pursuing the last grain of rice' one factor that may have assisted the manufacturing manager in the Japanese factory was the lack of traditional pressure from accountants. The company had one financial accountant and no management accountants; the manufacturing manager did not see any accounting data. However, he was required to report daily to Japan on physical output, efficiency and quality levels achieved.

An interesting issue is to what degree was the presence of Japanese managers vital to the maintenance and success of these various practices? The Japanese factory had a Japanese managing director and company secretary, and each UK line manager had a Japanese counterpart on the site. It was not clear how responsibilities were split between the UK and Japanese managers. It had been reported that when Japanese managers are withdrawn, standards begin to slip[14]. It was impossible to confirm this during this study, but the presence of Japanese managers may indicate that the Japanese parent was worried that this might happen.

A study of just two companies has many limitations as the companies may not be representative of the population of Japanese companies and UK companies trying to adopt Japanese practices. In addition, the practices may be transferable to the industries and the process technologies of the two companies studied but not to other industries. The following is a summary of the conclusions reached:

- The pattern of JIT manufacturing management practices used by a Japanese company in the UK is consistent with Hayes' description of practices used.
- Practices discounted or not considered by Hayes (in particular quality circles and robots) were adopted or considered by the UK company but not the Japanese company.
- The Japanese company had successfully transferred virtually all of these practices to the UK.
- The UK company had successfully adopted some of the practices.
- In general, Japanese JIT manufacturing management practices, as described by Hayes, are transferable to the UK. Some of these practices may require the presence of Japanese managers to ensure their successful application.

References

[1] Sugimori, Y., Kusnohi, K., Cho, F. and Uchikawa, S. 1977. Toyota production system and Kanban system: Materialisation of Just-in-time and respect for human system. *International Journal of Production Research*, 15(6): 553–564.

[2] Arbose, J.R. 1980. Quality control circles: The West adopts a Japanese concept. *International Management*, December 1980.

[3] Cole, R.E. 1980. Will QC circles work in the US? *Quality Progress*, July 1980.

[4] Lorenz, C. 1981. Learning from the Japanese. *Financial Times*, 1981.

[5] Rice, J.W. and Yoshikawa, T. 1982. A comparison of Kanban and MRP concepts for the control of repetitive manufacturing systems. *Production and Inventory Management*. First Quarter, 1982.

[6] Hatvaney, N. and Pucik, V. 1981. Japanese management practices and productivity. *Organisational Dynamics*, Spring 1981.

[7] Ouchi, W. and Jaeger, A. 1978. Type Z organisation: Stability in the midst of mobility. *Academy of Management Review*, April 1978.

[8] Reitsberger, W. The secret of Japan's success: Plain managerial competence. *Financial Times*.

[9] Hasegawa, Y. 1979. New developments in the field of industrial robots. *International Journal of Production Research*, 17(5): 447–454.

[10] Hayes, R. 1981. Why Japanese factories work. *Harvard Business Review*, July–August 1981.

[11] Hikichi, T. 1981. Japan's commitment to automation. Paper presented at BPICS Conference, *Productivity the Japanese Formula*. London, 20 February 1981.

[12] Magaziner, I.C. and Hout, T.M. 1980. *Japanese Industrial Policy*. Policy Studies Institute.

[13] Schonberger, R.J. 1982. *Japanese Manufacturing Techniques*. The Free Press, New York.

[14] White, M. 1982. *Japanese Management and British Workers*. Policy Studies Institute.
[15] Wheelwright, S.C. 1981. Japan – Where operations are really strategic. *Harvard Business Review*. July–August 1981.
[16] Mori, M. 1981. Future trends in Japan. Paper presented at BPICS Conference, *Productivity the Japanese Formula*. London, 20 February 1981.

2

MANUFACTURING TECHNIQUES

This section considers the core techniques of JIT in manufacturing. These are in three main areas: cellular manufacturing, Kanban (pull scheduling) and set-up time reduction. JIT focuses on balance and flow and cellular manufacturing is one of the key techniques supporting this. Group technology is an analytical technique leading to cellular manufacturing. To achieve JIT production, cellular manufacturing needs to be supported by small batch sizes. This can be brought about by the reduction of set-up times. The best known JIT technique is pull scheduling, normally known as the Kanban system. One paper in this section describes both the technique and its major variants, another paper on Kanban describes its application.

CELLULAR MANUFACTURING SYSTEMS

J. T. Black
Auburn University, USA

This paper presents an overview of this relatively new system – the cellular manufacturing system – and how it differs from the more conventional systems in use. Some guidelines for implementing a conversion to the cellular system are also given, and some of the major impediments encountered in such a conversion are detailed.

A new type of manufacturing system has emerged in recent years. It offers both cost and quality control advantages over the four traditional manufacturing systems, but it has yet to be implemented on any sizeable scale in the USA.

Traditional systems

Four kinds of manufacturing systems have traditionally been employed (Fig. 1). The oldest of these is the job shop, a transformation process in which units for different orders follow different paths or sequences through processes or machines.

The major characteristics of the job shop are flexibility, variety, highly skilled people, much indirect labour and a great deal of manual material handling (loading, unloading, setting up and adjusting). General purpose machines are grouped by function and adapted to the special requirements of different orders. The price for this flexibility is paid in the form of long in-process times, large in-process inventories, lost orders and poor quality.

For products built in larger quantities, the flow shop system can be used. The flow shop is a transformation process in which successive units of output undergo the same sequence of operations with more specialised equipment, usually involving a production line of some sort. No back flow is allowed.

Typically, all the units follow the sequence of operations and pass through all machines. Volumes are large, runs are long, and conversions to another product take a long time. The most automated

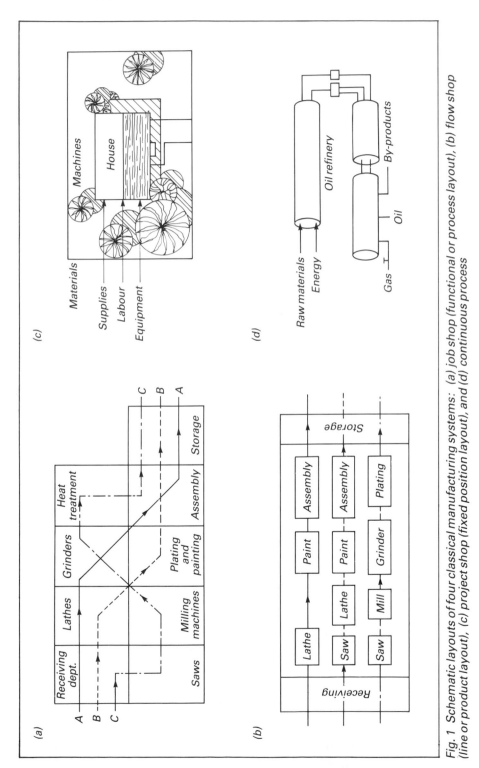

Fig. 1 Schematic layouts of four classical manufacturing systems: (a) job shop (functional or process layout), (b) flow shop (line or product layout), (c) project shop (fixed position layout), and (d) continuous process

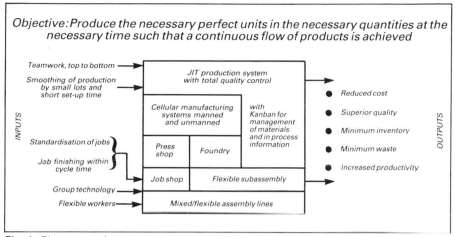

Fig. 2 Elements of the Just-in-time production system

examples of this sytem for machining are called transfer lines.

The third kind of system, the project shop, is directed toward creating a product or service which is either very large (immobile) or one of a kind with a set of well-defined tasks that typically must be accomplished in some specified sequence. The people, materials and machines all come to the project site for assembly and processing. The project is generally backed up by a job-shop/flow-shop system to supply component parts and subassemblies to the project.

The fourth type of manufacturing system is the continuous process system, in which products – generally gases, liquids or slurries – flow through a series of directly connected processes or operations that link raw material (inputs) with finished products (outputs).

Since this system does not deal with discrete parts, it is generally ignored by the non-chemical community. However, it represents the ideal, or *vision*, of the japanese Just-in-time production system (Fig. 2). Discrete products will ideally flow like water through the system.

The key to making this goal a reality on the plant floor is designing the manufacturing processes and systems such that *small lots* can be produced. The ideal 'lot' size is one. Small lots smooth the production flow.

This requires a fifth type of manufacturing system called the 'cellular manufacturing system' (also referred to as the group technology (GT) or work cell), see, for example, Fig. 3. The cellular system groups processes, people and machines to treat a specific group of parts. The output typically is completed components.

Tables 1 and 2 summarise the characteristics of all these manufacturing systems and give some examples of each kind. Each of the classical manufacturing systems has a mechanism for handling information and material movement and storage (inventory), purchasing, planning,

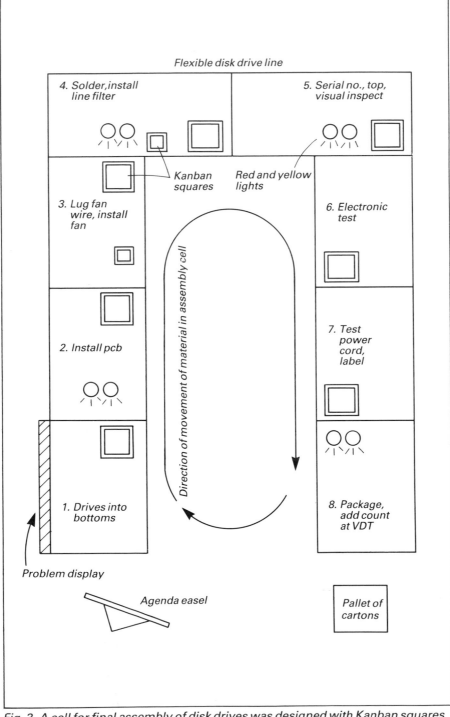

Fig. 3 A cell for final assembly of disk drives was designed with Kanban squares at Hewlett-Packard

Table 1 Characteristics of basic manufacturing systems

Characteristics	Job shop	Flow shop	Project shop	Continuous process
Types of machines	Flexible; general purpose	Special purpose; single functions	General purpose; mobile	Special purpose
Design of processes	Functional or process	Product flow layout	Project or fixed position layout	Product
Set-up time	Long, variable	Long	Variable	Very long
Workers	Single functioned, highly skilled (1 man to 1 machine)	One function; lower skilled	Single function skilled; 1 man to 1 machine	Few
Inventories	Large inventory to provide for large variety	Large to provide buffer storage	Variable, usually raw materials	Small in-process
Lot sizes	Small to medium	Large lot	Small lot	Not applicable
Production time per unit	Long, variable	Short; constant	Long; variable	Short; constant
Examples in goods industry	Machine shop; tool and die shop	TV factory; auto assembly line	Shipbuilding; house construction	Oil or chemical processing
Examples in service industry	Hospital; restaurant	College registration; cafeteria	Movie; TV show; play; buffet	Movie; TV show; play; buffet

Table 2 Characteristics of cellular and flexible manufacturing systems

Cellular manufacturing systems	Flexible manufacturing systems
• Small- to medium- sized lots of families of parts (1–200). A special set of parts or products • 1–15 machines • Rapid changeover – 'single set-up'[1] • One-at-a-time part movement within the cell • Defect prevention through integrated quality control	• Medium-sized lots of families of parts (200–10,000). • 5–12 CNC machines • Rapid changeover; no set-up at machines • Significant reductions in inventory • Greatly improved quality control through in-process inspection

Unmanned:
• Flexible/programmable machines (CNC)
• Robotic integration for parts handling (1–5 machines)
• Network computer control (see Fig. 11)
• Decouplers needed for flexibility

Manned:
• A set or group of general purpose machines and equipment laid out in a specific area
• Multifunctional workers
• Enhanced worker input leading to job enlargement
• Job enrichment
• Machine tools capable of completing cycle imitated by man
• Significant reduction in inventory between cells

(right column, Flexible manufacturing systems)
• Flexible/programmable machines
• Integrated conveyor system for parts and tooling
• Networked computer control

production control, and so forth (Fig. 4). Collectively, these activities represent the *production system*. The production system is designed around the manufacturing system, to support it.

The job shop represents the most common type of manufacturing system, however, with 30–50% of systems estimated to be of this form. In most job shops, some items will be made in large enough quantity to warrant the use of flow-line methods; therefore, it has been common practice in the USA to mix job-and flow-shop elements.

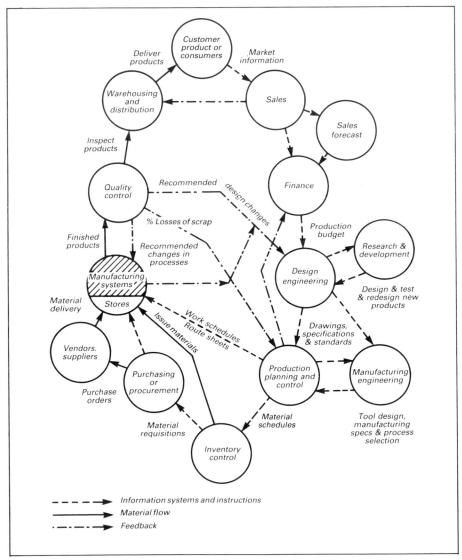

Fig. 4 Functions and systems in a production system as related to the basic manufacturing system (shaded circle) which may have any of the layouts shown in Figs. 1 and 3

Cellular manufacturing systems

The emergence of the cellular manufacturing system, and its highly automated form – the flexible manufacturing system (FMS) – has resulted in a new type of production system which is capable of producing high quality products at low cost. The best example of this new production system appears to be the Toyota Motor Company's Just-in-time (JIT) production system.

Group technology is a systems based rationale for solving the reorganisation problems involved in setting up cellular manufacturing systems. It provides a computer-oriented database and tools the manufacturing engineer can use to design the workcell. A GT analysis develops the families of parts which can be manufactured by a flexible, cellular grouping of machines. The machines in the cells can be retooled so that one can rapidly change from one lot of components to another, eliminating set-up time or reducing it to a matter of minutes.

Eliminating set-up time dramatically alters the economics of lot or batch production to permit the economical production of very small lots.

In equation form, TC (total cost) = FC (fixed cost) + VC (variable cost) × Q (quantity). Set-up is a fixed cost. It does not vary with the quantity made. Labour and mateial are variable costs. Fig. 5 is a graphic picture of this equation.

To obtain the cost per unit, divide the total cost equation by the quantity (Q): TC/Q = FC/Q + V. The new picture is shown in Fig. 6. Note what FC/Q represents. The fixed costs must be spread out over many units to reduce the cost per unit enough to make production of the item economical.

But what if you eliminate or greatly reduce the fixed cost by eliminating the set-up time and its related labour cost? Then the picture looks like Fig. 7.

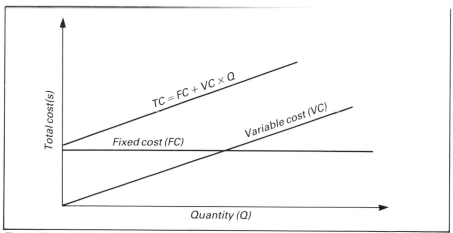

Fig. 5 Economics of lot or batch production

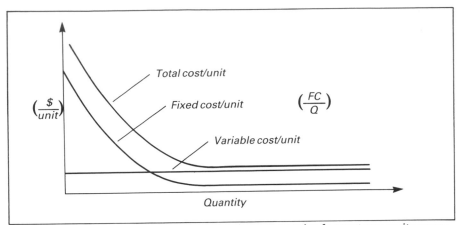

Fig.6 Classical representation of production economics for cost-per-unit versus quantity

Now it becomes as economical to make things in small lots as it is to make them in large lots.

You might ask, "How do you eliminate set-up? Isn't it a necessary evil?" It is not, but it takes work to eliminate it[1]. The truth is that American manufacturing never saw the need to reduce or eliminate set-up – they have operated on the basis of economic order and production quantities (EOQs and EPQs).

The Japanese have relegated EPQs and EOQs to the archives with their JIT/TQC (total quality control) system, because a primary objective of JIT is to reduce the lot size to the smallest size possible.

In cellular arrangements, one worker can hand a part directly to the next worker for another operation. If the part is defective, the process is halted to find out what went wrong. Quality feedback is *immediate*, and high quality products emerge. This *is* integration of the function of quality control into the manufacturing system.

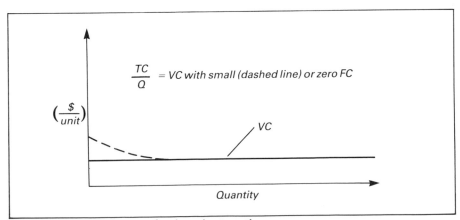

Fig.7 Effect on unit cost of reduced set-up time

Small lot quantity, coupled with a 100% perfect product (another hard-sought ideal), smoothes the production flow.

After many years and much hard work, the discrete part system begins to look more and more like a *continuous flow process* in which products flow like water through the plant. But the *key* first step to transforming a production system based on job-shop/flow-shop manufacturing systems is the transformation to a cellular manufacturing system.

Designing cellular systems

Cellular manufacturing has existed for many years, but it has not been properly defined or well understood, and it certainly hasn't been recognised as a particular type of manufacturing system. Let us define a manufacturing cell as a cluster or collection of machines designed and arranged to produce a specific group of component parts.

Few rules, and virtually no theory, exists for designing cellular manufacturing systems. However, the first rule is that the design should be as flexible as possible so that it can readily expand to include other components or be modified to handle additional members of the family. The objective is to link the cells into a large manned or unmanned integrated manufacturing system. Cells can be categorised into two general groups: *manned* and *unmanned*.

Manned cells

Manned cells contain machine tools which are conventional or programmable (NC or CNC machines) and production workers who have been trained and are skilled in the operation of more than one piece of equipment within the cell. The multifunctional worker is unusual in the typical job shop, but not in micro-electronics job shops, where workers have been extensively cross-trained with no difficulty. Manned cells are efficient because the number of workers can be adjusted and minimised to meet the desired output.

Regarding the design of manned cells, Monden[2] notes that the U-shape appears to offer the greatest flexibility because the range of jobs which the multifunctional worker can cover can easily be increased or decreased as needed. Typical manned cells are shown in Figs. 3 and 8.

Clearly, what these manned cells do is revoke Parkinson's Law. As the production requirements of the cell are reduced, the number of workers is reduced accordingly, and workers cannot simply expand the jobs to fill the increased time available.

The cells are linked to each other by a material handling system (JIT used Kanban[3]), or are directly linked, as shown in Fig. 9.

The author calls this kind of manufacturing system an integrated manufacturing production system (IMPS) because the functions which have traditionally been performed in the production system are now

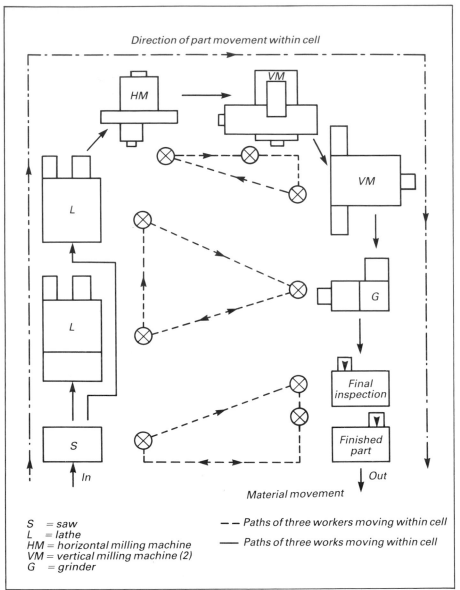

Direction of part movement within cell

VM

HM

VM

L

L

G

Final
inspection

Finished
part

S

In

Out

Material movement

S   = saw
L   = lathe
HM = horizontal milling machine
VM = vertical milling machine (2)
G   = grinder

— — Paths of three workers moving within cell

——— Paths of three works moving within cell

Fig.8 Schematic of a manned cell using conventional machine tools – cell is laid
out in U-shape and staffed by three multifunctional workers

performed in the manufacturing system. In the functionally designed
system (Fig. 9a), the milling machines are grouped separately from the
lathes. The production system which services this functionalised
manufacturing system (i.e. the job shop) is also organised functionally.
All the design engineers are in one place, all the quality engineers in
another place, and the production planners somewhere else; no
functional integration.

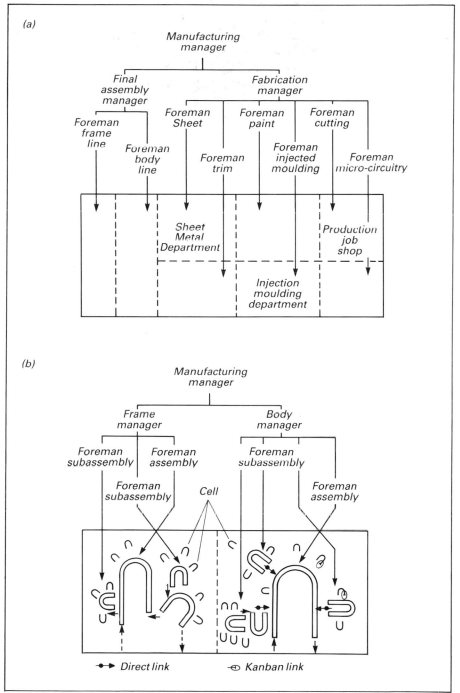

Fig.9 (a) The layout of a plant designed as a job is by process, so a functionalised layout and organisation result. (b) A plant designed with CMSs is product-oriented [2]. The cells are linked directly to the subassemblies (subprocesses put at point of use) or with Kanban.

In IMPS (Fig. 9b), production functions are integrated directly into the manufacturing system. Only the final assembly line is scheduled. Robots, computers and automation are added to solve problems. This system's key design feature is flexibility – the system can react quickly to changes in product demand and product design. Layouts which result in waiting times for the worker, large in-process inventories, isolation of the workers and situations in which worker waiting time is absorbed in producing unnecessary inventory, are avoided.

Linear manned layouts in which the worker walks from one machine to the next avoid most of these problems, but are not as flexible (in terms of rebalancing the number of workers in the cell in the event of demand changes) as the U-shaped cell.

In the manned cellular system, the worker is decoupled from the machines, so that the utility of the worker is no longer tied to the utility of the machine. (This means that there may be fewer workers in the cell than there are machines.) The objective is to improve the utilisation of the people by making them multifunctional, capable of running all the machines in the cell.

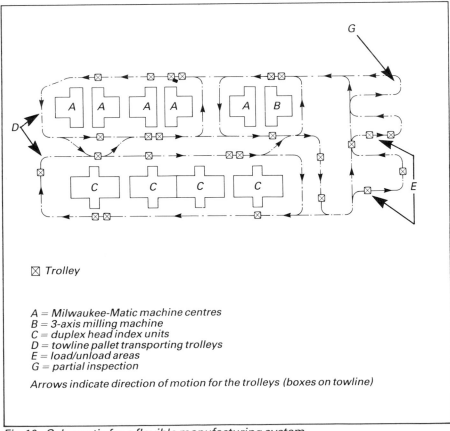

⊠ *Trolley*

A = Milwaukee-Matic machine centres
B = 3-axis milling machine
C = duplex head index units
D = towline pallet transporting trolleys
E = load/unload areas
G = partial inspection

Arrows indicate direction of motion for the trolleys (boxes on towline)

Fig.10 Schematic for a flexible manufacturing system

Unmanned cells

Unmanned cells contain machine tools that are programmable (CNC machine tools or other automated equipment), and there are few if any workers within the cell. Unmanned cells have a number of classes or arrangements. These are:

• *Fixed automated:* These cells are classically represented by the transfer line, in which the quantities are large (large lots) and the runs long. Such systems are generally arranged in lines, circles or the U-shape. They usually have a conveyor which both locates the part and transports it from the machine station, and the line is balanced such that the part spends the same amount of time at each station. The volume of parts is very large and the variety very small. These cells are not very flexible.

• *Flexible automated:* These cells are represented by the FMS (flexible manufacturing system) and the robotic cell.

The FMS is generally arranged in a line or a rectangular design with a computer-controlled conveyor to transport the parts to any machine in any order. The machines are programmable and therefore can change tools and machining programs to handle different parts (Fig. 10).

Parts can be introduced into the system in any order. Therefore, this system works on a family of component parts with medium to large lot sizes. These systems tend to be rather large and expensive, typically containing five to 12 machine tools; their cost parallels that of the transfer lines.

The robotic cell generally has few machines and is arranged so that a robot can load and unload the machines and change the tools in them, if necessary. In both the FMS and robotic designs, there should be liberal use of autonomation (automatic inspection) to ensure a high percentage of good parts.

Robotic cells are typically circular in design to take advantage of the range of motion of the robot (assuming the robot has a spherical or circular spatial range), but are not limited to such arrangements. As rectangular and mobile robots become more common, other designs will emerge.

In fact, as robots become more versatile, they will be able to communicate better with each other and will be able to hand parts to each other just as workers in manned cells hand parts to each other.

In most robotic cells in place today, there is only one robot. If there are multiple robots, they are placed on the floor so that they cannot reach each other or are programmed so that they cannot enter the same space at the same time.

When we have robots which are mobile within cells or can interact with each other, it appears likely that the FMS design will be employed only when parts are either too large (too heavy) for robots to pass from one to another or too small to be properly handled.

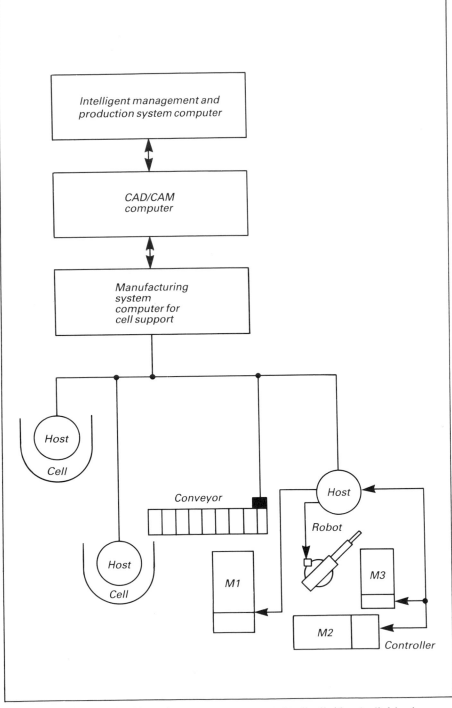

Fig.11 A computer (cell host) manages unmanned cells, linking individual
elements with higher-level control/information systems

The bottom line in these designs, however, is flexibility, with small lot sizes. Fig. 11 shows how the unmanned cell in a 'linked cell' manufacturing system can communicate with higher level computer systems.

Designing set-ups

In set-up design, the objective is to eliminate set-ups between the different component parts in the parts family or, at worst, to be able to use essentially the same fixturing for every part in the family with quick action modification, or 'one-touch set-up,' between parts. In manned cells, this objective can be readily accomplished – as has been demonstrated by numerous Japanese and American manufacturers – with the result that one can economically manufacture in very small lots.

Furthermore, the concept of single set-up, which means that set-up is accomplished in less than ten minutes, can be extended to presswork and foundry areas which are still essentially in lot-type shop production. In the press shop at Toyota, for example, workers routinely change dies in the presses in three to five minutes or less. (The same job at many American or European companies may take four to five hours.) This has the effect of markedly reducing inventory turns. However, the Japanese have found that the *main* benefits are superior quality, worker motivation and enhanced productivity.

Unfortunately, there is no information on how to achieve single set-up in any standard American text on manufacturing or tool engineering design. Very little research has been done on the elimination of set-up time in the USA. Why is this? Because we have never seen the need.

Need for new approach

The job shop is, generally speaking, the least productive manufacturing system. It results in products or services which are costly and whose costs tend to keep rising with inflation. In addition, some social and technological trends suggest that the number of and need for small lot production systems will increase in the future. These trends or needs include:

• Proliferation of numbers and varieties of products. This results in smaller production lot sizes (as variety is increased) and decreased product life-cycles. Set-up cost, as a percentage of total cost, becomes greater, as do the problems of managing inventories and materials.
• Greater need for closer tolerances and greater precision.
• Increased cost of product and service *liability* as consumers demand accountability on the part of manufacturers, which requires greater emphasis on reliability and quality.

● Increased variety in materials, with more diverse properties, which requires great flexibility and diversity in the processes used to machine or form these new materials.

● Increased cost of energy needed to transform materials and of capital and materials.

● The need to markedly improve productivity and reduce the costs of goods and services to halt inflation and meet international competition from those who are using these systems.

● The move away from a labour-intensive manufacturing environment toward a service environment.

● The need for shorter service times or lead-times in production to reduce inventories and allow faster response to changes in demand.

● Worker demands for improved quality of working life.

Overlaying these trends is the continued rapid growth of process technology, led by computer technology. No one can forecast the ultimate magnitude of this technological development, but it is clear that no segment of manufacturing and production will escape its impact. Computer aided design (CAD), manufacturing (CAM), process planning (CAPP), testing and inspection (CATI), and integrated information systems are rapidly being integrated into the factory of tomorrow.

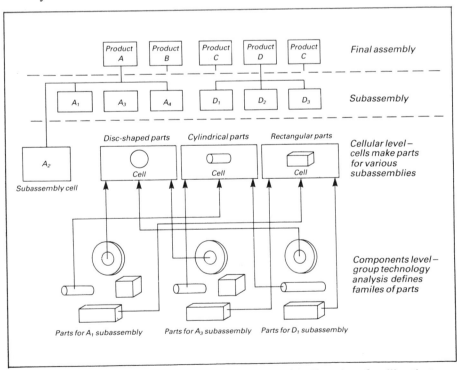

Fig.12 *Grouping of components from various product lines into families that can be fabricated in cells*

A group technology programme

One of the best ways to reorganise a system is by embarking on a group technology or GT programme. Since its initiation in the USSR in the late 1950s[4], the group technology concept has been carried throughout the industrialised world. It is now well rooted in Germany, the USSR, the UK and, especially, Japan, where it is a 'way of life' in many manufacturing facilities.

GT provides a systems approach to the redesign and reorganisation of the functional shop. It will have an impact on every segment of the existing system. It is a manufacturing management philosophy which *identifies and exploits the 'sameness' of items and processes used in manufacturing industries.*

GT groups units or components into families of parts which have similar design or manufacturing sequences (Fig. 12). Machines are then collected into groups or cells (machine cells) to process the family. Portions of the functional system are converted, in steps, to the cellular system or the flexible manufacturing system.

This implies a number of things. It means redesigning the entire production system and all functions related to it. The change will affect product design, tool design and engineering, production scheduling and control, inventories and their control, purchasing, quality control and inspection, and of course the production worker, the foreman, the supervisor, the middle manager and so on, right up to top management. Such a conversion must be viewed as a *long-term transformation* from one type of production system to another.

It is unlikely that the entire shop can be converted into families, even in the long run. Therefore, the total collection of manufacturing systems will be a mix that evolves toward that ideal of a continuous process system over the years. This will create scheduling problems, because in-process times for components will be vastly different for products made by cellular or flexible systems and those made under traditional job shop conditions.

Finding families of parts is one of the first steps in converting the functional system to a cellular/flexible system. There are many ways to do this, but three popular ways are through:

- Tacit judgement, or eyeballing.
- Analysis of the production flow.
- Coding and classification.

The eyeball method is, of course, the easiest and least expensive, but also the least comprehensive. This technique clearly works for restaurants where 'chicken' represents a family of parts, and for large lots of similar parts, but not in large job shops where the number of components may approach 5000–10,000 and the number of machines may be 300–500.

The second method, product flow analysis (PFA)[5], uses the information available on route cards.

The idea, which is illustrated in Fig. 13 is to sort through all the components and group them by matrix analysis, using route sheet information to develop the initial matrix. This method is more analytical than tacit judgment, but not quite as comprehensive as the coding/classification method.

PFA can get rather cumbersome when the number of components and/or machines becomes large. Sampling can be used (20–30%) to alleviate the size problem, but then it is uncertain whether all the potential members of the family have been identified and how big the family really is. However, PFA is a valuable tool for use in systems reorganisation. Part of this technique involves analysing the flow of materials in the entire factory to lay the groundwork for the new plant layout.

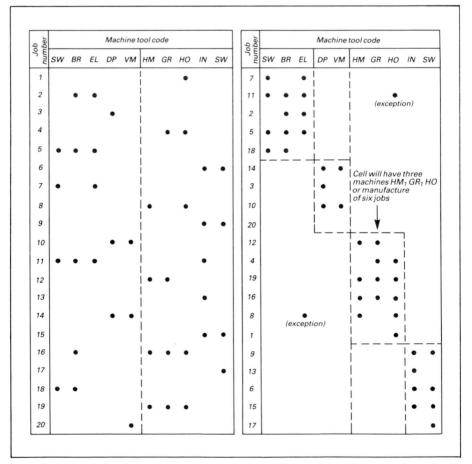

Fig. 13 Matrix (a) of jobs (by number) and machine tools (by code letter) as found in the typical job shop, and (b) rearranged by PFA to yield families of parts and associated groups of machines that form a cell

The overwhelming majority of companies that have converted to a cellular system have used a coding/classification (C/C) method. There are design codes, manufacturing codes and codes that cover both design and manufacture.

Classification uses coding to sort items into classes or families based on their similarities. Coding is the assignment of symbols (letters or numbers or both) to specific component elements based on differences in shape, function, material, size, process, etc.

Numerous C/C systems have been published[6], and many have been developed by consulting firms. Most systems are computer compatible, so that computer sorting of the codes generates the classes or families. The system developed at John Deere and Co. in Moline, Illinois, is perhaps the most usable but refined GT system in the world..

Whatever C/C system is selected it should be tailored to the production of the particular company and should be as simple as possible, so that everyone understands it. It is not necessary that old part numbers be discarded, but every component will have to be coded prior to the next step in the program, finding families of parts. This coding procedure will be costly and time consuming, but most companies opting for this conversion understand the necessity of performing this analysis.

The composition of parts families, and therefore of the cells that are designed to manufacture them, is dependent upon what characteristics you decide to sort according to. Family generation after coding and classification is not automatic.

Forming cells and FMS

Depending upon their material flow, different families will require different layout designs. In other words, the manner in which the family is determined influences the design of the cell. In some families, every part will go to every machine in exactly the same sequence; no machine will be skipped, and no back flow will be allowed. This is, of course, the purest form of a cellular system, except perhaps for the single machining centre in which all parts are made on one machine.

Other families may require that not all components go to all machines, or that the forward sequence of order through the machines be modifiable, or that back flow in the group be provided for. Flexible manufacturing systems are designs to accommodate these situations, as are many cellular systems. This in no way alters the basic concepts, but it does add to the complexity of scheduling parts through the machines.

The formation of families of parts leads to the design of the cell, but this step is by no means automatic. Design is the critical step in the reorganisation and must be carefully planned. Remember, the objective here is to convert the functional system, with its functional layout, into a *flexible* group layout.

We are just beginning to come to grips with the problems of cellular design (see Table 3) and how to make these cells truly flexible, but the following guidelines for the formation of machine groups have proved useful:

● There should be a sufficient volume of work in the family to justify establishment of a cell. Remember, however, that the capacity of the cell (its output rate) is determined by the number of workers in the cell as dictated by the needs of the manufacturing system. Some families may be too small to form a group of machines, but would be good candidates for processing on a single CNC or machine centre, which is the smallest form of a cell.

● The composition of a component family should permit a satisfactory worker utilisation situation. In manned cells, this may mean that not every machine will be utilised 100%, or even that the machine utilisation rate will be greater than it was in the functional system

The objective in manned cellular manufacturing is to improve the utilisation of the *people*. In designing its manufacturing/production systems, Japan's chief objective is to make them flexible, and this should be our objective as well. In unmanned cells, machine utilisation is obviously an important consideration.

● The processes in these systems should be technologically compatible.

● The needed capacity of the system must be determined from the quantities of parts needed and the production schedule, which determines *when* they are needed. Problems involving balancing labour and machine utilisation will be encountered.

● The physical reorganisation of the manufacturing systems will necessitate the redesign of the production system. The problems outlined below will have to be addressed.

Product design will be impacted – in terms of both new parts and standardisation of old parts – but the designers will be able to see parts being made in total in the cells and to get a better grasp of the manufacturing system and what it can make.

Planning and scheduling for cellular designs will be different. Scheduling of a mixed shop will involve difficulties with respect to timing components to arrive at assembly points on time. However, scheduling will be easier overall, because the control has moved on to where the work is being done – in the cell itself.

Some people argue that manned cellular processing can be accomplished by simply routing families through machines without ever forming them into cells. This defeats many of the benefits of the cellular system and will obviously never lead to conversion of the functional system to a cellular or flexible system. For example, workers in a cell

Table 3 How to form cells

Integrated manufacturing production systems are based on a linked cellular manufacturing system design. Knowing how to design the cells to be flexible is the key to successful manufacturing.

- *Make tacit judgements based on axiomatic design principles:*
  - *minimise functional requirements*
  - *simplify the design of system*
  - *minimise the design information in the product*
  - *decouple those elements which are functionally coupled*
  - *integrate production system functions into the manufacturing system*
- *Use group technology methodology:*
  - *production flow analysis finds families and defines cells*
  - *coding/classification is more complete and expensive*
  - *other GT methods*
- *Simulation:*
  - *digital simulation of the system*
  - *physical simulation of the system*
  - *object-oriented and graphical simulation*
- *Pick a product or products:*
  - *design a linked-cell manufacturing system beginning with the final assembly line (convert FA to mixed model) and move backward through subassembly to component parts and suppliers*
- *Eyeball techniques:*
  - *find the key machine, often a machining centre, and declare all parts going to this machine a family; move machines needed to complete all parts in family around key machine*
  - *build cell around a common set of components like gears, splines, spindles, rotors, hubs, shafts, etc.*
  - *build cell around common set of processes; example being drill, bore, ream, keyseat, chamfer holes*
  - *build cell around set of parts which eliminates the longest (most time consuming) element in set-ups between parts being made in the cell*

Cells are linked together using Kanban or directly

tend to become more flexible because they become familiar with the entire family and can see the common elements in all the parts in the family.

The manned cellular system provides the worker with a natural environment for job enlargement. Greater job involvement enhances job enrichment possibilities and clearly provides an ideal arrangement for improving quality.

In advanced forms of unmanned cellular layouts, the micro computers of the CNC machine tools are networked together with a robot for material handling within a cell. It is difficult, if not impossible, to conceive of this kind of arrangement without some method of collecting the work into compatible families. All the machines in the cell are programmable, and therefore this kind of automation is very flexible as well as more economical than traditional job-shop processing.

Benefits of conversion

Conversion to these new forms of manufacturing/production systems can result in significant cost savings over a two to three year period and marked improvements in quality.

However, reorganisation has a greater, though immeasurable, benefit. It prepares the soil so that the 'seeds' of the computer aided and automated manufacturing of the future will fall on fertile ground. The progression from the functional shop to the shop with manned cells to clusters of CNC machines to an entire system of 'linked cells' must be accomplished in logical, economically justified steps, each building from the previous state. Simply adding robots and computers to your existing job shop will not make you as efficient as your competitor who has successfully undertaken such a system conversion.

Constraints on implementation

Clearly, such a conversion requires a major effort. Some of the constraints on the implementation of a cellular conversion programme are as follows:

● Systems changes are inherently difficult and costly to implement. Changing the entire manufacturing production system is a huge job.
● Companies are willing to spend freely for product innovation, but not for process innovation. Few machine tool companies sell integrated and coordinated systems with their machining centres.
● Decision making is choosing among the options in the face of uncertainty. The greater the uncertainty, the more likely the 'do nothing' option is to be selected. This is because of fear of change and the unknown.
● Many companies use faulty criteria for decision making. Decisions for change should be based on the company's ability to compete (quality, reliability, delivery time, flexibility for product change or volume change), not on the basis of output (production rate) or cost alone. The high cost plus the long-term payback on such conversions equals a high risk situation in the minds of the decision maker. (But is there really an alternative?)
● Short-time financially oriented viewpoints conflict with the long-term nature of programme.
● Unions' fear of loss of jobs and resistance to the multifunctional worker concept can act as deterrents.
● The general lack of blue collar involvement in the decision making process in companies leads to poor vertical communication.
● Because the greatest impact of such conversions is on middle management, this group generally provides the greatest resistance to change.

Concluding remarks

Converting to integrated manufacturing production systems is a systems level change. This means that it requires careful planning and the full cooperation of everyone involved. All must understand the problems, costs and limitations of making such a conversion and the long-term effort that will be needed.

Table 4 Aliases for integrated manufacturing production systems

- Just-in-time/total quality control (Schonberger's terminology – most popular in USA)
- ZIPS (zero inventory production system used by Omark Industries)
- MAN (material as needed, term for IMPS used by Harley Davidson)
- MIPS (minimum inventory production system coined by Westinghouse)
- Ohno System (after Toyota's Taiichi Ohno, mastermind of the system and used by many companies in Japan)
- Toyota production system (the 'model' in reality)
- Stockless production (Hewlett-Packard – a misnomer since no process can run without stock)
- Kanban – many companies in USA, Japan

However, the magnitude of this change also means that it offers the potential of tremendous savings and meaningful advantages. The integrated manufacturing system (or its aliases listed in Table 4), with its simple aim of making goods flow through a plant like the waters of a river, has amply demonstrated the scope of the benefits to be gained.

References

[1] Shingo, S. 1983. *A Revolution in Manufacturing: The SMED System*. Productivity Press, Cambridge, Massachusetts.
[2] Monden, Y. 1983. *Toyota Production System*. Industrial Engineering and Management Press, IIE, Atlanta.
[3] Schonberger, R.J. 1983. Plant layout becomes product-oriented with cellular, Just-in-time production concepts. *Industrial Engineering*, 15(11): 66-71.
[4] Mitrofanov, S.P. 1977. Scientific Organization of Batch Production. AFML/LTV Technical Report TR-77-218, Vol. III. Wright Patterson AFB, Ohio.
[5] Burbridge, J.L. 1979. *Group Technology in the Engineering Industry*. Mechanical Engineering Publications, London.

CELLULAR MANUFACTURING AT JOHN DEERE

W.J. Dumolien and W.P. Santen
Deere & Co., USA

Cellular manufacturing at Deere & Co. has grown from its 1975
introduction at the component works facility in Waterloo, Iowa, to
become a major management philosophy. Cellular manufacturing is
the transfer of raw material into subassemblies or finished parts within
a single organisational entity, or a cell. Deere has implemented a large
number of manufacturing cells containing from 10 to 30 machine tools
producing several thousand part numbers. The introduction of these
cellular manufacturing systems has coincided with facility
reorganisation and modernisation activities. The practice of cellular
manufacturing has evolved into one of John Deere's most important
manufacturing strategies.

To reorganise functional manufacturing into a cellular arrangement,
the part population must be divided into groups that can be processed
as completely as possible within a single cell. Parts are segregated into
families. Families are matched with machines to form cells, and the
cells are balanced to achieve acceptable machine and operator
utilisation.

Tacit judgement or manual analysis alone is not feasible for
large-scale projects. A computerised 'systems approach' is required to
review the vast amounts of necessary data. The John Deere group
technology system was developed to support the design of cellular
manufacturing systems.

This system is a collection of integrated programs for data capture,
coding, planning and design. The heart of the system is a classification
and coding scheme developed by John Deere engineers for the capture
of geometric characteristics of parts.

Data are entered through an interactive classification program which
generates a code that references geometric part features. The system
also contains production data such as part operation routings, current
and future production requirements, cost amounts and machine load
information.

All data are available to the engineer through a set of standardised on-line computer modules. These include:

- Code – provides part data entry.
- Extract – isolates a group of parts using desired criteria.
- Analysis – provides a set of analytical routines such as machine loads and part flows.
- Modify – allows changes to data to perform 'what if' analyses.
- File – provides file handling programs and interface with statistical analysis programs.

The system is a decision support tool which allows the creation, comparison and evaluation of cellular manufacturing alternatives. The creation of these manufacturing cells is typically approached using the standard phases of analysis, detailed design, implementation and operation.

Data analysis

Establishment of the part database containing the related geometry and production information becomes the first of three cellular analysis steps. Part numbers, processes, equipment, machine utilisation and operating procedures must be determined.

Part families are formed from this part population. Each family is a collection of parts with the same machining features as determined from the coded geometry. Since the code represents the geometric features resulting from the transformation of raw material into finished parts, all parts with the same intrinsic operational requirements are grouped together. This method is used to avoid the proliferation of less than optimal routings rather than grouping according to routing information.

In the second analysis step, part families are grouped into cells for routing evaluation based on one of three themes of cell development. The most general grouping theme is similar geometric features.

The optimum routings are determined for families of geometrically similar parts. The machines needed in these routings become candidates for a cellular design. An example of this may include families of rectangular, flat bar parts with machined slots of similar size and shape, produced as complete within one dedicated machine cell.

A second theme used in cell formation is based on similar process sequencing. Families with similar sequential routings can be assigned to cells which contain the machine tools needed to perform only those operations. Part families with sequential operations of milling, drilling and induction hardening may be candidates for this type of cell formation.

The third theme of cell formation focuses on product lines and subassemblies. These cells include all operations, both machine and assembly, needed to produce completed subassemblies or products. In

Table 1 GT cell department operation

	Department No.							
	1	*2*	*3*	*4*	*5*	*6*	*7*	*8*
Number of parts	77	107	55	59	57	58	75	103
Number of families	12	20	5	6	7	8	8	9
Number of cells	2	4	2	2	2	3	1	3
Operations completed within department (%)	99.6	91.8	94.7	95.5	99.5	99.9	94.0	100.0
Number of operations	455	442	398	536	694	667	421	497
Number of unique processes	6	6	6	6	5	4	6	5

evaluating a varied part population, all these themes of cell formation may be applied.

The third analysis step is the calculation of the machine utilisation for each proposed cell. Cells with relatively high machine utilisation will be implemented. Cells with low machine usage must be combined with other cells. Cells with requirements for the same capital intensive equipment are combined to share this equipment. This grouping method provides for effective utilisation of expensive equipment.

When combinations of cells will not improve machine utilisation, the routing of part families within the cells is revised. Parts are re-routed to similar machines for improved utilisation and cell integrity.

Generally, machines capable of producing the larger sized parts are selected as key machines in re-routing the problem part families. This assures adequate machine capacity, but it may cause a shortage of large size machine tools.

It is also necessary to organise the manufacturing cells into departments using the same grouping principles. The objective is to complete all operations, if possible, within one department and maximise machine utilisation.

Department size criteria must include the effects of part requirement fluctuation as well as standard manageability factors. Examples of proposed departments are shown in Table 1.

Detailed design

The previous analysis developed the group technology plan of action for cellular development. The detailed design phase must address the specifics of machine layout, common tooling and operator requirements. Also included are evaluations of operational improvements and the introduction of new technology.

The specific machine tools needed in each cell must be laid out.

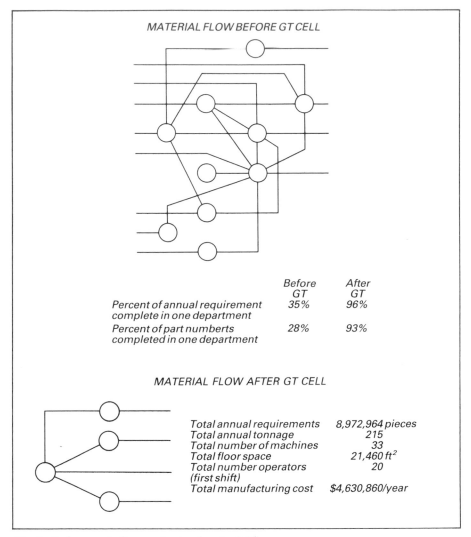

Fig. 1   Before and after analysis of material flow

Machine tool performance must match the requirements of each cell. The machine tools are arranged to minimise material handling and eliminate backward flow through the cell. An example of streamlined material flow is described schematically in Fig. 1.

One of the basic benefits of cellular manufacturing is the reduction of tooling set-ups from the use of common part family tooling. Tooling requirements are reviewed and substitutions made to minimise the number of required tool changes.

For example, a family of eight spindle parts which now use common tooling are permanently affixed to a shared machine tool. Previously these parts were produced on five different machines with eight tool changes. Set-ups, tool crib investments and job change costs are all reduced.

Cellular manning methods are established for effective use of human resources. It becomes advantageous to have operators capable of running multiple machine tools. This provides a greater degree of flexibility within a cell. The operators are now assigned to a group of finished parts instead of to individual operations.

For example, a flexible cell may include four lathes and four hobbing machines run by three operators. While an operator performed a job change on one machine, the other operators would maintain production on the remaining seven machines.

Once the cellular design has been approved, implementation follows basic project management techniques. Physically relocating machine tools is a minor effort compared to the extensive staff manpower requirements associated with a cellular arrangement. Tooling records, mechanical data and job detail standards must be revised. Without these revisions, start-up of machine cells will be prolonged and unnecessarily difficult.

The detailed design and implementation phases of cellular manufacturing projects require significantly more effort than the analysis phase. This is exemplified in the manpower percentages shown in Table 2. The project involved the creation of 19 cells producing 600 part numbers.

Table 2 Manpower requirements for GT cell formation

Requirements	Percent of total time
Analysis	
Initial cellular groupings	4
Design	
Process planning	18
Process and tool	17
Industrial engineering – methods and standards	14
Preset tooling	14
Manufacturing engineering services	14
Implementation	
Facilities redevelopment	13
Plat engineering	5
Material engineering	1

The cellular core

At the core of cellular manufacturing processes are the machines, tools and floor space required for each operation. Early attempts at designing cells concentrated effort on the establishment of this cell core.

It is now realised that cellular manufacturing practices offer productivity improvements which are shared between staff and shop-floor activities. Cellular manufacturing is part of an overall management strategy. All affected employees should be versed in the cellular concept and the focus and simplification of manufacturing that it provides.

Product engineering can make use of family design practices when introducing new parts. Parts of similar function, when possible, can be designed according to a family standard.

This effort helps to ensure the design integrity and manufacturability of new part introductions. These new parts can then be incorporated into existing manufacturing cells using standard routings.

Process planning and methods engineering must realise the advantages of utilising part family production. The focus and reduction of routing alternatives simplifies process planning activities. By design, optimal family routings are determined within manufacturing cells. Therefore, many new parts that fall into existing part families will have the optimum routing predetermined.

Process planning and methods engineering efforts are greatly reduced. The prediction of new product manufacturing cost will also be more accurate due to the focused production scheme of cellular manufacturing.

Advantages to production control include part family scheduling. When possible, part families, instead of individual part numbers, should be scheduled for production. Parts should be scheduled within a family before changing to other family groups.

Part families which make use of similar tooling increase productivity by requiring fewer set-ups.

Quality of finished parts is improved due to the increased visibility of part processing operations. In most cases, one small group of operators is responsible for the complete production of finished parts in sequential operations.

In these cells, quality problems are easily seen by the operators responsible and can be corrected immediately. Fewer inspectors and increased part quality are potential benefits of cellular manufacturing systems.

It must be realised that many staff-supported areas are affected in the development and execution of manufacturing cells. There is no 'turnkey' method for the design and execution of cellular manufacturing.

Each cell can have characteristics dependent on the depth of planned and monitored benefits. Only with strong management direction will the potential savings of a cellular manufacturing strategy be realised.

Results

Based on the success of cellular manufacturing systems, the application of group technology concepts has increased dramatically since their introduction at John Deere. Cellular manufacturing is a recognised strategy for improved productivity.

Studies of other manufacturing techniques, particularly the relationship of cells to the Just-in-time philosophy, have further demonstrated the importance of cellular manufacturing as a cost-effective management strategy.

The actual benefits of the completed cellular manufacturing systems at three John Deere manufacturing facilities have validated the predicted benefits. Typically, the results are:

- A 25% reduction in the number of required machine tools.
- A 70% reduction in the number of departments responsible for the manufacture of a part.
- A 56% reduction in job change and material handling.
- An 8-to-1 reduction in required lead-times and a corresponding reduction of inventory.
- Shop supervisors who now have more control over processing, with clear delineation of responsibility.

Since their introduction, group technology and cellular manufacturing concepts have been developed, refined, promoted and practiced. Today, all new production and facility improvements at John Deere are analysed using the group technology system for the application of cellular manufacturing. Current projects at separate facilities are examining cellular manufacturing for cast iron and steel machining, chassis production, subassembly welding and metal forming.

The past success of the implemented manufacturing cells and the continued expansion of these concepts demonstrate the importance of cellular manufacturing at John Deere.

THE KANBAN SYSTEM

R.J. Schonberger
Schonberger and Associates Inc., USA

Kanban is a Japanese technique for requesting more manufacturing
materials on an as-needed basis by issuing a kanban (card), plus an
empty container. Kanban is a pull system of replenishment, as
opposed to the push system that governs schedule-based systems like
material requirements planning (MRP). Toyota's dual-card kanban
technique, with a productivity improvement feature, is contrasted with
a less-potent and simpler technique employing a single kanban.
Kanban techniques are also compared with MRP, reorder point, and
continuous replenishment techniques.

Kanban (pronounced kahn-bahn), literally translated, means visible
record or visible plate. More generally, kanban is taken to mean card.
The Toyota Kanban system employs a card to signal the need to *deliver*
more parts and an identical or similar card to signal the need to *produce*
more parts.

If the Kanban system is very loosely interpreted to mean any system
employing an order card or delivery card, then most companies could
claim to have one. It has long been standard procedure in industry for a
card of some kind to accompany work in process; a 'traveller'. And a
variety of cards or forms – job orders, route sheets, job tickets, and so
forth – are commonly used in ordering more parts. These do *not*
constitute a Kanban system, because they are part of a *push* system of
parts ordering and control. A distinctive feature of the Toyota Kanban
system is that it is a *pull* system.

Push or pull?

For the past 18 years or so the American Production and Inventory
Control Society has provided workshops, expert speakers, and training
materials that have found their way into about every American hamlet
that has a manufacturing company. Included in the message is the view
that a well-planned computer-based *push* system of manufacturing
planning and control is a key to effectiveness.

In reality, a push is simply a schedule-based system. That is, a multi-period master production schedule of future demands for the company's products is prepared, and the computer explodes that schedule into detailed schedules for making or buying the component parts. It is a push system in that the schedule pushes the production people into making the required parts and pushing the parts out and onward. The name given to this push system is material requirements planning (MRP).

In the old days before we had the computer power to do all this planning and scheduling, a haphazard pull system was used (and still is in a good many companies). It works as follows: customers place orders, and manufacturing looks to see if the parts are on hand. Parts not on hand are pulled through, or expedited. Even if substantial amounts of parts are kept on hand, there will be a few missing ones that must be expedited; this is disruptive and keeps customers waiting.

A push schedule, or MRP, system seems like good management when compared to a pull or expedite system. But a weakness of MRP is that there is some guesswork involved. You need to guess what customer demand will be in order to prepare the schedule, and you need to guess how long it will take your production department to make the needed parts. The system allows corrections to be made daily (called shop-floor control). Nevertheless bad guesses result in excess inventories of some parts, though not nearly so much total inventory as in the old pull or expedite system.

Until recently, it appeared that pull systems would gradually be forced out of existence by computer-based MRP, even in small companies given the low and still-falling cost of microcomputers. But the Toyota pull system, known as Kanban, upsets that prediction. Kanban provides parts when they are needed but without guesswork and therefore without the excess inventory that results from bad guesses. But there is an important limitation to the use of Kanban. Kanban will work well only in the context of a Just-in-time (JIT) system in general, and the set-up time/lot size reduction feature of JIT in particular[1]. A JIT programme can succeed without a Kanban subsystem, but Kanban makes no sense independent of JIT.

The Toyota Kanban system

In the Toyota Kanban system every component part type, or part number, has its own special container designed to hold a precise quantity of the part number, preferably a very small quantity. There are two cards, or 'kanban', for each container, and the kanban identify the part number, container capacity, and certain other information. One kanban, the production kanban, serves the workcentre producing the part number; the other called a conveyance kanban, serves the workcentre using it[2]. Each container cycles from the producing workcentre and its stock point to the using workcentre and its stock point, and back, and one kanban is exchanged for the other along the way (Fig. 1).

Rules

The Kanban system's simplicity and effectiveness are intertwined in the following rules:

- No parts may be made in milling unless there is a P-kanban authorising it. Milling comes to a halt rather than make parts not yet asked for – a pure pull system. (Workers may do maintenance or work on improvement projects when there are no P-kanbans in the dispatch box.)
- There is precisely one C-kanban and one P-kanban for each container, and the number of containers per part number in the system is a carefully considered management decision.
- Only standard containers may be used, and they are always filled with the prescribed (small) quantity – no more, no less. With such careful control of quantities per container, as well as number of containers per part number, inventory control is simple and far more precise than manual or computer-based Western systems.

Kanban as a productivity improvement system

The second rule states that the number of containers is a carefully considered management decision. Too many containers is too much inventory in the system. In Western thinking too few containers is too little inventory in the system; not so in the Toyota system, which employs the extraordinary Japanese concept of deliberately removing buffer inventory (or labour) in order to experience – and solve – problems[3,4]. Kanban is ideal for removing buffers. The foreman needs only to remove kanban from the system. Removing kanban is sufficient, because an empty container without kanban attached is ignored and gathers dust. As an illustration of the effects of kanban removal, we may return to the example of milling and drilling (Fig. 1).

Assuming that the process is stable and that there are five kanbans in the system; that means five C-kanbans, five P-kanbans and five containers of milled heads. Now the shop foreman over both workcentres cuts the inventory to four kanbans. The likely effect is that milling will experience its normal problems, and at bad times of the day will have trouble keeping up with drilling. For example, in a certain two-hour period, milling might find that some of its newly milled heads do not meet specifications, perhaps because of worn bearings in a milling machine, or because of tool wear; a minor accident might send a machinist to the dispensary for first aid; a machine might break down; small variations in dimensions of the heads to be milled might cause set-up delays.

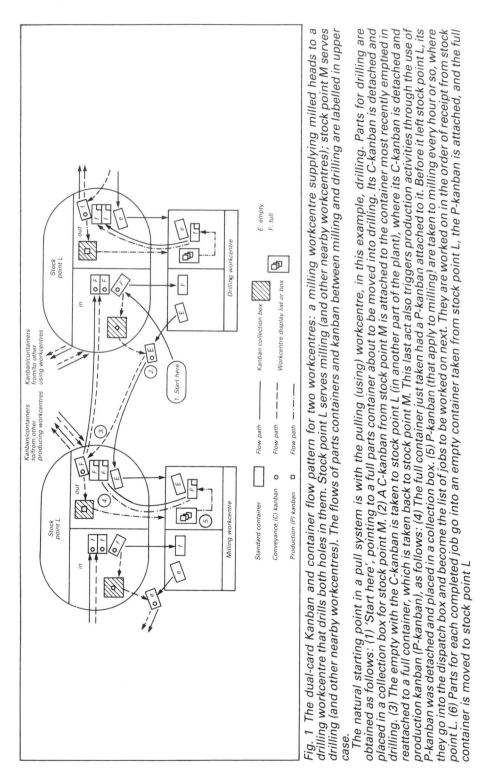

Fig. 1 The dual-card Kanban and container flow pattern for two workcentres: a milling workcentre supplying milled heads to a drilling workcentre that drills both holes in them. Stock point L serves milling (and other nearby workcentres); stock point M serves drilling (and other nearby workcentres). The flows of parts containers and kanban between milling and drilling are labelled in upper case.

The natural starting point in a pull system is with the pulling (using) workcentre, in this example, drilling. Parts for drilling are obtained as follows: (1) 'Start here', pointing to a full parts container about to be moved into drilling. Its C-kanban is detached and placed in a collection box for stock point M. (2) A C-kanban from stock point M is attached to the container most recently emptied in drilling. (3) The empty with the C-kanban is taken to stock point L (in another part of the plant), where its C-kanban is detached and reattached to a full container, which is taken back to stock point M. This last act also triggers production activities through the use of production kanban (P-kanban), as follows: (4) The full container just taken had a P-kanban attached to it. Before it left stock point L, its P-kanban was detached and placed in a collection box. (5) P-kanban (that apply to milling) are taken to milling every hour or so, where they go into the dispatch box and become the list of jobs to be worked on next. They are worked on in the order of receipt from stock point L. (6) Parts for each completed job go into an empty container taken from stock point L, the P-kanban is attached, and the full container is moved to stock point L

Such events slow down milling's rate of output, perhaps enough so that drilling uses up three full containers of heads and is idled while milling completes the order to fill the fourth container. At the end of the day both drilling and milling might be behind schedule, which is apparent in two ways:

- P-kanbans and empty containers for certain models have piled up – not a good way to start up production the next day.
- A count of the day's production – perhaps just a simple total of tally marks on a paper or blackboard, where one tally signifies one kanban of heads milled – reveals underproduction. That is, the daily schedule, expressed either by model or in total of all models, has not been met. For these kinds of parts – milled heads and drilled heads – it is likely that the daily schedule for making the parts exactly matches the daily schedule for the manufacturer's end product, which might be engines. Usually, the foreman would direct the two workcentres to work overtime until the day's schedule is met, but the P-kanbans supersede the schedule as the authorisation to produce.

Workers, group leaders, and the foreman are not pleased about failing to meet the schedule, and most would rather not have unplanned overtime thrust on them. They are, on the other hand, pleased to have unearthed a new set of problems to attack. In the Kanban and Just-in-time system, workers are always gathering data on the next set of problems, and periodically they are showered with praise when a problem is solved. To earn praise, to avert criticism, to gain self-satisfaction, and to avoid unplanned overtime, Kanban workers generally are supportive and enthusiastic about the productivity improvement features of Kanban.

Of course, the causes of the problems unearthed must be carefully recorded for later analysis by the group, who may be given some company time on certain days of the week for improvement projects. In the above example of milling problems, some possible solutions, stopgap measures, and corrective actions might include:

- Seeking management approval to establish a formal project team to study the problem of milled heads not meeting specifications.
- Asking maintenance to investigate machine breakdowns and providing them with the latest breakdown data.
- Placing a first-aid kit in the shop for very minor injuries.
- Getting quality control involved in the problem of small variations in dimensions of the heads, so that the problem may be traced to its source.

Kanban limitations

Kanban is feasible in almost any plant that makes goods in whole (discrete) units (but not in the process industries). It is beneficial only in certain circumstances:

- Kanban should be an element of a JIT system. A pull system makes little sense if it takes interminably long to pull the necessary parts from the producing workcentre, as would be thc case if set-up times took hours or lot sizes were large. The central feature of JIT is cutting set-up times and lot sizes, which allows for fast 'pulls' of parts from producing work centres.
- The parts included in the Kanban system should be used every day. Kanban provides for at least one full container of a given part number to be on hand all the time, which is not much inventory idleness if the full container is used up the same day it is produced. (Toyota's rule-of-thumb is that the maximum capacity of containers for one part number should not exceed 1/10 a day's usage[1].) Therefore, companies with a Kanban system generally apply it only to the high-use part numbers, but replenish low-use items by conventional Western techniques (e.g. MRP or reorder point).
- Very expensive or very large items should not be included in Kanban. Such items are costly to store and carry. Therefore their ordering and delivery should be regulated very closely under the watchful eye of a planner or buyer.

There are numerous fine points to the Toyota dual-card Kanban system. The interested reader will find several sources in English that explain Toyota Kanban elsewhere[1,2,4]. What has not been reported on is a popular Kanban simplification, single-card Kanban.

Single-card Kanban

Only a small number of Japanese companies have implemented the complete Toyota dual-card Kanban system. Yet probably hundreds claim to have a Kanban system. What most have is a single-card Kanban system, and the single card that they use is a conveyance kanban (C-kanban). It is easy to begin with a C-kanban system, and then add P-kanban later if it seems beneficial.

In single-card Kanban, parts are produced and bought according to a daily schedule, and deliveries to the user are controlled by C-kanban. In effect, the single-card system is a push system for production coupled with a pull system for deliveries (Fig. 2).

Single-card Kanban does not employ a stock point for incoming parts. Instead, parts are delivered right to the point of use in drilling. Also, the stock point for parts just produced tends to be larger than that for dual-card Kanban. The reason for the enlarged stock point is that it holds stock produced to a schedule; the schedule pushes milled parts into the stock point even when drilling has been slowed or halted as a result of production or quality problems. So the stock point must be able to hold more containers of parts than in the pull system.

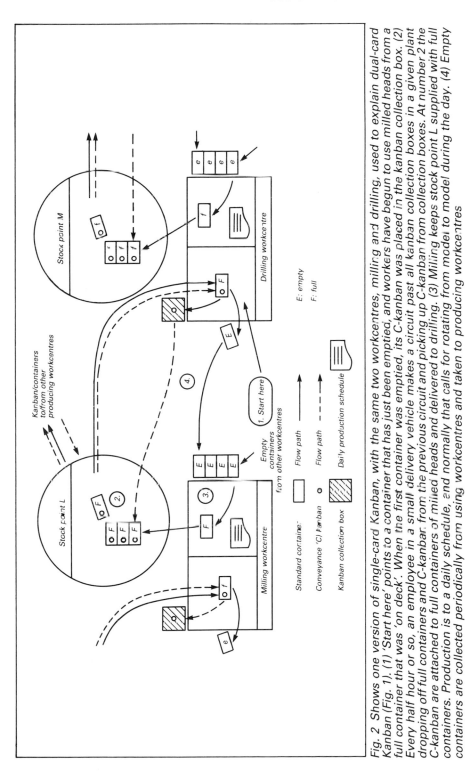

Fig. 2 Shows one version of single-card Kanban, with the same two workcentres, milling and drilling, used to explain dual-card Kanban (Fig. 1). (1) 'Start here' points to a container that has just been emptied, and workers have begun to use milled heads from a full container that was 'on deck'. When the first container was emptied, its C-kanban was placed in the kanban collection box. (2) Every half hour or so, an employee in a small delivery vehicle makes a circuit past all kanban collection boxes in a given plant dropping off full containers and C-kanban from the previous circuit and picking up C-kanban from collection boxes. At number 2 the C-kanban are attached to full containers of milled heads and delivered to drilling. (3) Milling keeps stock point L supplied with full containers, and normally that calls for rotating from model to model during the day. (4) Empty containers are collected periodically from using workcentres and taken to producing workcentres

Single-card compared with dual-card Kanban

Single-card Kanban controls deliveries very tightly, so that the using workcentre never has more than a container or two of parts, and the stock point serving the using workcentre is eliminated. An advantage is that clutter and confusion around points of use is relieved.

On the other hand, produced parts may be allowed to build up to some excess in the stock point serving the producing workcentre. But the build-up need not be serious in companies in which it is relatively easy to associate the quantity and timing of component parts with the schedule of end products. There are many examples: motorcycles, motors, pumps, generators, consumer appliances, and toys.

Compare, for example, a motorcycle plant with an automobile plant. Perhaps the motorcycle plant makes eight sizes of motorcycles in three colours, and type A always takes a type A frame, type A engine, type A fenders, and so forth; there are very few customer options. Therefore, if the assembly schedule calls for 20 type As to be completed per day, then 20 fenders per day are needed at, say, minus one hour (one hour prior to completion of each assembly); 20 engines and frames per day are needed at perhaps minus two hours; and so forth. Tube pieces for frames might be punched out at minus one and a half days, but since the tubing must then go through frame welding and painting, where problems and delays can occur, the schedule might better be set at minus two days. The extra half day provides small buffer stocks of tube pieces, completed frames, and painted frames. Single-card Kanban nicely controls deliveries of parts from one stage to the next, and the daily parts schedules, appropriately offset for lead time, provide the parts when needed, with rather small inventory buildups.

The automobile plant would have perhaps ten times as many parts, many colours, hundreds of customer choices, and many more stages of production. Compared with motorcycles, there is a far greater potential for delay, a compound effect of: (a) large numbers of parts, (b) variable occurrence factors, and (c) multiple stages of manufacture. Daily schedules for producing each part number would have to provide for sizeable buffer stocks (through extreme backscheduling) in order to avoid running out of parts when the delays are bad. Toyota's ingenious solution to the problem is dual-card Kanban, which signals production of each part number to match the up-and-down output rate of succeeding production stages.

Dual-card Kanban is doubly effective in that it has the productivity improvement feature of removing Kanban to expose and solve problems. Unfortunately, single-card Kanban cannot employ that feature, because there is no control on number of full containers of a given part number. Therefore, companies that use single-card Kanban must get their productivity improvements in some other way. For example, Kawasaki Motors, Corp., a single-card Kanban company, gets productivity improvements by removing workers from final assembly until yellow lights come on signifying problems in need of correction[3]. Nihon Radiator Company, also a single-card Kanban user, has a vigorous total quality control system, which features a continual succession of improvement projects; the improvement projects that deal with quality, work methods, tools or equipment improve productivity by reducing material and labour per unit and by improving equipment and tool utilisation[3].

Is single-card Kanban unique?

Among American managers who tour single-card Kanban plants, there is invariably someone who offers the perceptive comment, 'it looks like the old two-bin system to me'. The two-bin system is a visual reorder point technique: when you see that the supply of a part number is down to where you must open the last box (or dip into the second bin), you reorder. Single-card Kanban does indeed work that way. It is unique only as an element in a Just-in-time system, including but not limited to the following:

- Standard containers are used.
- The quantity per container is exact, so that inventory is easy to count and control.
- The number of full containers at the point of use is only one or two.
- The quantity in the container is small so that at least one container and usually several containers are used up daily.
- At the producing end, the containers are filled in small lot sizes, which requires prior action to cut set-up times and thereby make small lots economical.

The C-kanban is merely an identification card and a convenient signal to bring more parts. There are other effective ways to signal the need for more. At one Kawasaki plant in Akashi, Japan, one workcentre conveys the same message by rolling a coloured golf ball down a pipe to the producing workcentre; the colour identifies which part number[1]. At Mitsuboshi Belting Company in Kobe, scannable computer cards serve as C-kanban; where more parts are needed an online scanning device may send a message to a producing workcentre (or to inventory control)[3]. A telephone or intercom could be used – or even a yell, if the producing workcentre is within earshot.

And, finally, the system could be modified slightly so that the empty container itself signals the need for more parts. Master Lock, an American company, has had such a container-signalling system for making more parts for its locks for over 35 years[5] (but without the five JIT attributes listed above). The necessary modification for a container-signalling system is strict control of the number of empty containers; extra empties cannot be sitting around out on the floor, or somebody will assume that they must be filled.

Whether to adopt dual- or single-card Kanban depends on a crucial factor, the *ease of associating parts requirements with the schedule of end products*. If it is extremely difficult to associate parts needs and end product schedules, then neither type of Kanban is very good. Well-known Western inventory systems may be better.

Kanban and other inventory systems

The oldest and most widely used inventory system in the world is the reorder point. The simple reorder point rule is: when stocks get low, order more. But reorder point (ROP) results in high inventories. Parts and raw materials are ordered because of the rule instead of need. Manufacturers that use ROP do so because of a difficulty in *associating parts requirements* with the *schedule of end products* – the crucial factor.

Material requirements planning (MRP), provides a better way. MRP harnesses the computer to perform thousands of simple calculations to transform a master schedule of end products into parts requirements. But MRP shares one weakness with ROP: it is lot oriented. That is, in the MRP process the computer collects all demands for a given part number in a given time period, and recommends production or purchase of the part number in one sizeable *lot*. MRP companies order in lots, rather than piece for piece (just-in-time), because American MRP companies have not lowered set-up times in order to make small lots economical. If they did so, then simple manual Kanban rather than complex, expensive computer-based MRP would be their logical choice. The MRP paradox is that, if the company removes the set-up time obstacle in order to make MRP truly effective in cutting inventories to the bone, then MRP is no longer needed; Kanban is preferable.

The point may be stated in terms of the 'crucial factor': MRP correctly calculates parts requirements by precisely associating them with the master schedule of end products. But what is correct at the time of calculation is subject to error later. The reason is that lots are sizeable, and the production lead time is long, from one to several weeks. During that lead time there will be delays and schedule changes so that the lot being produced no longer is correct for the master schedule of end products. The lot size and lead time erode the close association between parts requirements and end-product schedules.

The four inventory systems discussed in this paper, single-card Kanban, dual-card Kanban, MRP, and ROP, seem to fall naturally onto a continuum (Fig. 3).

Weeks' or months' worth of inventory is typical of the reorder point (ROP) system (Type 5), the dominant system for companies with large delays (lead time) between parts manufacture and parts use. Some companies, especially small job shops, deliberately position themselves in a market where customers are buying variety and the manufacturer's flexibility. They cannot schedule in advance and therefore resort to high inventories replenished via ROP. If products are rather complex (multiple stages of manufacture), inventories will build further because of buffer stock between stages.

Days' or weeks' worth of materials is typical for the MRP companies (Type 4 in Fig. 3), mostly found in the USA. These producers are able to shuck off the high inventory ROP system, because they have a closer link between parts and end-product demands, which makes future scheduling of end products realistic. Such companies may sell fewer models (less variety) than the Type 5 companies and to a market that demands faster deliveries; but production is in lots, and a produced lot often covers demands for a week or more.

Days' or hours' worth of materials is typical for the Toyota dual-card Kanban system (Type 3 in Fig. 3). Further reduction in number of models, plus engineering to cut set-up times, reduces the delay from parts manufacture to parts use – and cuts lot sizes to approach piece-for-piece production.

Companies using single-card Kanban (Type 2), like Kawasaki and Nihon Radiator, have still less delay or lead time between parts production and use. Such companies have very few models, few levels of manufacture, few required raw materials, or have made a concerted attack on sources of delay, and perhaps other reasons. The manufacturing process has been so simplified that inventory could be cut to perhaps minutes' worth, but hours' worth is more typical.

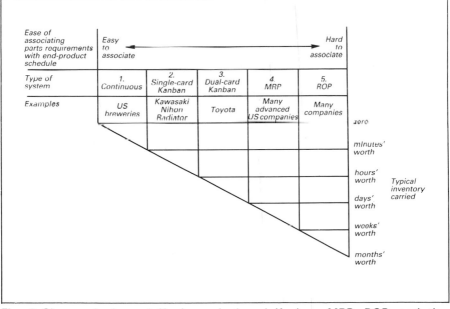

Fig. 3 Shows single-card Kanban, dual-card Kanban, MRP, ROP, and the continuous system, in a continuum. As it becomes harder to associate parts and end-product demands, inventories are likely to increase – from theoretical zero on the extreme left to months' worth on the extreme right

Companies that have fully streamlined plants with each production line dedicated to one model, produce a flow of goods like water without stops for inventory build-up, so that the minimum of zero inventory is imaginable (Type 1 in Fig. 3). US breweries are not quite continuous producers since the liquid goes into cans, kegs and bottles (discrete units). But the lead time from empty to full can is very short so that the canning schedule may nearly serve as the delivery schedule for incoming empties. By one report[6], Anheuser-Busch unloads empties from a nearly continuous stream of trucks, and uses them so soon that the average inventory of unfilled cans is only two hours.

This is referred to as a 'continuous' system, which in this case refers to the nature of the flow of materials and not the system for controlling the flow. Actually, when a company enjoys that degree of streamlined flow, just about any inventory system will keep inventories low – even reorder point.

At Matsuhita Electric's highly automated refrigerator plant, reorder point is used[3]; inventories are low but could probably be even lower if there were much savings involved. But bigger savings lie elsewhere. Type 1 manufacturers need not worry about inventories much, and Type 2 firms are somewhat similar, so their inventory systems are usually simple. Type 3 and 4 companies have huge potential inventories to be concerned about, and their inventory systems therefore receive great attention. Type 5 companies have the greatest inventory problems, and indeed they often fail because of inability to finance the inventory burden. They must either develop into Type 4 or 3 companies, or stay healthy by providing unique products or services.

Is Western industry ready for Kanban?

The answer is, usually not! Usually Western companies do not have conditions that make Kanban useful. Western brewers, bottlers, and canners are too much like the process industry to use Kanban; Kanban is a technique for discrete, not continuous, processes. Western companies using reorder point or material requirements planning could adopt single- or dual-card Kanban, but they would not gain any Just-in-time benefits in doing so.

Before implementing Kanban, lot sizes and buffer stocks should be cut. Reducing quality-related delays, stoppages, and rework will also reduce the lead time and bring the demand for parts into closer synchronisation with demand for end products. After such preparations, some companies may find it natural to adopt single-card Kanban. MRP might be retained for scheduling (the push system), and C-kanban used to pull the parts from producing to using workcentres.

The author knows of two American-owned plants that have adopted a Kanban procedure of this kind. The Greeley Division of Hewlett-Packard Corp. has installed a C-kanban to pull parts from a stockroom – parts previously scheduled by MRP. In the Cleveland area one

General Electric plant is supplying parts to another GE plant on roughly a daily basis, with Kanban triggering the deliveries.

Some MRP companies, especially those with multiple levels of processing, may evolve to Type 3 companies – resembling Toyota in regard to ease of associating parts demand with end-product demand schedules. In that case, MRP could be retained for capacity planning purposes, and dual-card Kanban adopted for pulling parts through the manufacturing processes. Yamaha has developed a half-way system that marries MRP to dual-card Kanban; Yamaha calls it *synchro-MRP*[1]. Synchro-MRP may be the best choice for some Western MRP companies able to streamline.

Finally, Western ROP companies that are able to streamline their processing and reduce the uncertainty of the coupling between parts and end-product schedules often will want to bypass the MRP stage. MRP is expensive, and it attacks problems with complex solutions, that is, computer systems. Where there are many stages of production, MRP or synchro-MRP may be necessary. In most cases, however, money is better spent on JIT than on computer-based planning and control. If we have learned anything from the Japanese, it is that simplification is generally the safest path to improvement.

References

[1] Hall, R.W. 1981, *Driving the Productivity Machine: Production Planning and Control in Japan.* American Production and Inventory Control Society, Falls Church, VA.
[2] Monden, Y. 1981. Adaptable kanban systems help Toyota maintain their just-in-time production. *Industrial Engineering*, 13(5): 29–46.
[3] Schonberger, R.J. 1982, *Japanese Manufacturing Techniques: Nine Hidden Lessons in Simplicity.* The Free Press, New York.
[4] Sugimori, Y., Kusunoki, K., Cho, F. and Uchikawa, S. 1977. Toyota production system and kanban system: Materialization of just-in-time and respect-for-human system. *International Journal of Production Research*, 15(6): 553–564.
[5] Hintz, R. 1981. Presentation to the Just-in-time workshop of the APICS repetitive manufacturing group.
[6] Eisen, G. 1981. Speech to Omaha chapter of National Purchasing Management Association.

SET-UP TIME REDUCTION
MAKING JIT WORK

D.L. Lee
Cummins Engine Co. Ltd, UK

This paper outlines the process adopted by Cummins Engine Company to meet the challenge of significantly reducing the duration of machine downtime involved in the set-up and changeover of equipment producing a variety of components. The programme utilises a basic work study approach, supplemented by a modern aid – the video camera. Cummin is one of the world's largest leading designers and producers of diesel engines. The company manufactures a range of over 100 models with outputs of 56–1800bhp. Cummins' engines power trucks and buses, construction equipment, rail cars and locomotives, marine craft and power generation plant. Over 4500 people are employed by Cummins in the UK, and the company exports almost 70% of its production from three modern British plants. The Daventry facility, home of the high horsepower K series diesels, manufactures engines for markets in Europe, Africa, the Middle East and the Far East.

Much is currently written and talked about the JIT concept of manufacturing. It probably is still the most misunderstood term in modern manufacturing language, despite the inclusion of JIT somewhere on the content outline of many current management conferences and seminars.

The intent of this paper is not to delve into the whole JIT philosophy, rather to deal with one of the fundamental needs of a JIT approach to manufacturing in a batch oriented machine shop – the elimination or significant reduction of the delays caused when equipment is stopped in order to set it up differently between the production of components of varying configurations.

The JIT philosophy

Let's examine the JIT concept briefly then, only sufficiently for us to be convinced of the need and merits of short set-ups.

Basically, those 'converted' to JIT believe that inventory is *evil* – it represents an indifference that we in the Western world have developed towards some of the problems we encounter in manufacturing:

- Equipment breakdowns.
- Unpredictable quality.
- Inflexible labour.
- Long set-up and changeover times.

What do we do when these things are ingredients in our day-to-day production routines? We put inventory around us in an attempt to protect ourselves from them – we begin to get cushioned from their effect and therefore the pressure to address them is removed, or at least lessened. We have invented unintentionally a way of ensuring that problems survive – reducing efficiency, increasing lead-times, lowering responsiveness and adding significant cost in the form of storage space, preservation, material handling, stock control/expediting systems, etc.

The JIT philosophy challenges this approach however. By demanding that we 'make only the *minimum* necessary *units* in the *smallest possible quantities* at the *latest possible time*' it requires that we identify and address all of the barriers that otherwise would prevent us from doing so.

Whilst some 'JIT experts' advise that inventory levels are progressively reduced in order to highlight some of the problems and allow us to attack them, others, rightly in the author's opinion, say that we should not risk hurting our customers while we go about increasing our competitiveness – we know the nature of many of the problems and should attack these first before taking risks by significantly reducing our inventories.

One such known problem is 'set-up' which exists wherever a piece of equipment is used to produce a variety of products of differing configuirations and if we can do something to reduce its duration then we can do it more frequently and without penalty make smaller batches – we are well on the way to eliminating some of the weaknesses we have already touched upon. Sounds easy, but where do we start?

The traditional engineering approach

As business manager of a production unit manufacturing the cylinder blocks for large, high horsepower engines of both 12 and 16 cylinder configurations the author asked his process engineer for his recommendations as to how set-ups could be reduced and what the likely cost would be.

Work flow analysis had already identified that of the 30 plus operations, a third required set-up between the two types of product. Duration was between 10 minutes and 90 minutes per machine, and totalled around 9.5 hours for the 10 machines.

The engineer calculated that a capital investment of £10,000 would initially remove 2.0 hours of the total (21% reduction) and a subsequent investment of £10,000 would get us another 0.75 hour reduction (10% of the new base). After that, as you would expect, things got tougher in compliance with the law of diminishing returns –

big money for small improvements. Fairly sophisticated machinery (mostly CNC) and fixturing (hydraulic clamping, fixed datums, etc.) are already used and the study confirmed the expectations that it would be costly to engineer the changes.

A modified approach

Attendance at an in-company seminar conducted in the USA by Ed Hay of the Rath & Strong Consultancy caused the author to change his views as to how the set-up reduction (SUR) programme should be approached.

Hay had spent several years in Japan, studying their approach to JIT and had since practised a 'Westernised' approach to the concept in the USA and Canada. The author modified still further Hay's recommendations in consultation with the plant's management group so as to ensure 'fit' with our own culture, and a plant strategy was developed as follows:

- Initially, the SUR programme would be implemented on the cylinder block production line (Fig. 1) – a line already charged with *delivering* its product Just-in-time to the assembly operation (i.e. from a small bank of finished machined components held at the end of the machining line) and with a need to convert speedily to a line capable of *manufacturing* its product in a JIT concept (utilising a 'one-at-a-time' Kanban approach).

- The author would develop and conduct, incrementally, an appropriate training programme for all block line production operators and immediate support staff. From the attendees, we would form SUR teams who would then spearhead the improvement programme.

Fig. 1 General view of a portion of the cylinder block manufacturing line

- The programme would be a *no cost/low cost* approach with limited funding available (expense budgets). Capital expenditure, if required later, would need to be fully justified via the normal capital approval process.

- The SUR teams would be *action* teams, as opposed to *study* teams with a target of 50% reduction in set-up duration after six months, and a further 25% reduction (of the start point) after operating for 12 months.

- The savings made in set-up duration would be re-invested in more frequent set-ups to produce correspondingly smaller batches, i.e. when duration was reduced to 50% of the starting figure, batches would be cut to 50% of their original quantity, and be produced twice as frequently. Hence the company would be nearer to producing components in the sequence consumed by the customer (engine assembly), would lower work in progress and hence lead times.

- The company would advocate to the SUR teams the use of video recording techniques (with in-built time recorder) as a means of recording both the current and ultimately developed methods of conducting set-up activity. This would require the support of the workers and their union representatives and the generation of trust, particularly as set-up activities had never before been subject to any form of analysis or time study, despite the existence of work measurement on 'standard' production operations.

The training programme

During the early part of 1986, a training package was developed, and incrementally conducted, commencing with the cylinder block line. Each manufacturing work team, along with supervisory, technical and other direct support staff spent approximately 10.5 hours (3 × 3.5 hour sessions) learning about:

- The JIT concept – an overview of its origins, benefits and the state of implementation in Japan and the West.
- JIT manufacturing – uniform plant load, line balancing, plant layout implications, Kanban systems.
- The changing business environment – what this means for the plant and the changes that people (the trainees) would see going on around them, particularly in the assembly business which had commenced a JIT assembly pilot project for the initial stages of engine build. This portion of the training led up to a recognition of the need to reduce set-up times.
- Team composition
- Recording and analysis methods.
- Course summary and 'next steps contract'.

On completion of the course each attendee received a certificate validating his successful completion and also a comprehensive reference manual containing a copy of all the course material, as well as analysis and improvement guide notes and associated forms.

The author believes the success of the programme, which commenced as each work area completed its training, was to a large extent due to the way in which the last three subject headings (above) were taught and subsequently executed. For that reason, each of them is expanded upon changing the emphasis from what was recommended and taught during the course, to what was subsequently installed as a process (there were few modifications thought necessary).

Team composition

When one observes the events taking place during a typical machine set-up, it becomes immediately apparent that much of the activity (or lack of it) is centred around factors which are not engineered in to the process and which therefore cannot be bought-out by large injections of capital/advanced technology. They are more likely to be the result of:

- The operator not knowing the next job.
- Blunt/inappropriate tools.
- Unclear specifications.
- Lost or damaged tools or equipment.
- Awkward access to work area or adjustment points (guards permanently welded or bolted because they once rattled or fell off.)

Because of these considerations and because the team is intended to be an *action*-oriented team, rather than a study group, we need the people closest to the action and therefore the real 'experts' as core members – *the operators who perform the set-up.*

During training two to three operators who were regularly involved in the set-up of the machine in focus were advocated. They were supported by technical people – in this case the line process engineer and the tool control supervisor (a person able to get minor improvements to fixtures and tools done quickly within his own functional area).

Operating experience of the teams dictated there was a need to add a maintenance operator to the core group because of the number of occasions on which problems were deemed to be caused by poor maintenance (or more correctly by poor communication between manufacturing and maintenance people).

The *facilitator* role was filled by an industrial (work study) engineer and revolved around calling meetings, issuing minutes and chairing brainstorming sessions, etc.

The author, as the trainer, also found a role in the team activities as a *guide* – it was thought sensible to have someone around, initially at least, to help the teams gain recognition and find their way through the myriad of organisations and systems which exist in any company.

It was decided before the training commenced (and supported by the views of the trainees later) that the line supervisor should not be a core member of the team for two important reasons:

- It would be unreasonable, given the extent of his already-stretched commitments to expect that he could further divide his time between (what would become) numerous SUR teams.
- There was a risk, because of the traditional supervisor–operator relationship that the supervisor could become the person who finished up with the majority of the team work assignments owing to his demonstrated skills in terms of 'getting things done'. This would overload him eventually and could cause the team to become a 'study' group who delegated work to other (already stretched) resources.

It therefore became vital, if the supervisor's support were to be won and kept, that he be updated regularly on the teams' activities – via minutes, occasional invitation to team meetings and, importantly, by constant updating by the team facilitator.

Recording and analysis methods

No better method exists than the use of a video camera to record what *actually* happens when a set-up takes place (Fig. 2). The video tape becomes a 'living' record – very much superior to either the written or spoken word which often tends to suggest what *ought*, rather than did

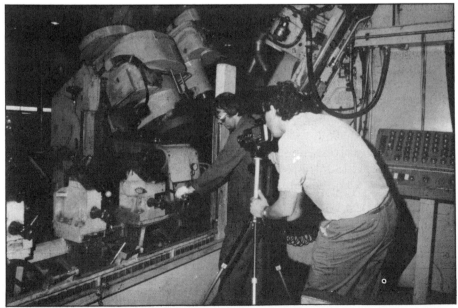

Fig. 2 Set-up activity being video recorded

occur. Better still is a video system which, as many now do, places a time-counter feature at the lower corner of the picture. Even someone who has conducted hundreds of time studies over the years has his eyes opened by the crudeness of some of the operations which we ask operators to perform daily, but faced with a video replay cannot conveniently code the activity IT (ineffective time) and lose it somewhere in the margin of a study sheet.

Two considerations of great importance emerge, however, when we plan to use video on the shop-floor. These are:

- That the recording is done openly and becomes the property of the team. There needs to be an environment that ensures that the team (particularly the shop-floor members) can trust beyond doubt that the video will not be misused in any way. Even for 'outsiders' to mildly scorn the way work is performed currently may be sufficient to jeopardise the whole programme and video is so searching that the uninformed may be tempted to make off-the-cuff derisory comments – *don't* – much of what is seen on video represents what *management* puts working people through daily.

- To video a set-up requires a small amount of preparation. The risk here is that with some advance notice we may witness an untypical set-up – we need it to be representative of what normally occurs. For this reason it was stressed throughout the training that, for instance, if it was not normal practice to collect tools and equipment in advance of the set-up, then this should not be performed for the sake of the camera. Nor should the camera be stopped if things go wrong (tight threads, jammed wheels, etc.) – we need to film it 'like it really is' – what really happens.

Soon after the video is made, the first team meeting takes place. The video is reviewed and an analysis sheet filled out (see Fig. 3). This could, alternatively be done for expediency outside of the main meeting. Work is broken into elements and classified by type and duration as follows:

C: Clamp
A: Adjustment
O: Other (describe)
UC: Unclamp
CD: Cleandown
P: Problem (describe)

With this completed, and a simple Pareto analysis conducted to rank in highest order the time consumers, the brainstorming can commence. Some early improvements can be identified easily if it is possible to safely move work from being internal, i.e. essential as part of the set-up stoppage, to external (can be done safely before or after with machine producing).

Cummins SET UP AND CHANGEOVER ANALYSIS SHEET		WORK TEAM BLOCKLINE 'C' WRITTEN BY S. KELCHER DATE 1/3/86 REVISION Nº		FROM COMPT. (DESC) KV 12 BLOCK NUMBER 3176466 TO COMPT. (DESC) KV 16 BLOCK NUMBER 3176468 MACHINE (DESC) ROUGH MILL BORER PLANT Nº 1510	
STEP NUMBER	WORK PERFORMED	WORK CATEGORY	TIME (MINUTES) FROM / TO / DURATION	SUGGESTIONS FOR IMPROVEMENT	RESP.
10	AIRBLOW CLAMPING BOLTS AND SLIDEWAYS ON BOTH HEADSTOCKS.	C D	00.00 / 2.00 / 2.00	ELIMINATE CAP HEAD BOLTS, FIT SWARF DEFLECTORS IF REQ^D.	I.S. P.L.
20	COLLECT HAND TOOLS	U C	2.00 / 2.20 / 0.20		
30	RELEASE CLAMPING BOLTS ON BOTH HEADSTOCKS	U C	2.20 / 5.45 / 3.25	IMPROVE CLAMPING METHODS	I.S. P.L.
40	CHANGE AIRLINE TO AIRGUN	P	5.45 / 6.15 / 0.30	INSTALL ANOTHER AIR SUPPLY	R.B.
50	DRIVE PIN MISSING, COLLAR LOOSE, USE ALLEN KEY INSTEAD	P	6.15 / 7.20 / 1.05	IMPROVE DESIGN OF COLLAR AND RETAINING PIN, MAKE SPARES.	I.S.
60	DRIVE LEADSCREW TO POSITION L.H. HEADSTOCK	A	7.20 / 10.25 / 3.05	CLEAN AND SERVICE LEADSCREWS FIT DEEPER SQUARE DRIVE SOCKET	O.LL. O.K. I.S. S.K.
70	FINAL POSITIONING USING OPEN ENDED SPANNER	A	10.25 / 10.40 / 0.15		

Fig. 3 Set-up and changeover analysis sheet

Other areas for improvement are:

Cleandown (CD): Provision of covers, caps, deflector plates, etc. to prevent material ingress to critical areas. (Example of guide notes, see Fig. 4).

Clamp (C): Review/reduce (with some care) unnecessary clamping points – engineers do seem to like 30:1 factors of safety; which in the event of a mishap can actually cause problems. Do away with inefficient screwthread fasteners – there are hundreds of better ways for frequently-moved fastening methods.

- *Are we cleaning down the appropriate areas as required to set-up, i.e. real set-up need or could it be done shift-end, etc.*
- *Can any of the necessary cleandown be performed safely before the machine is stopped for set-up*
- *Can any awkward corners or pockets be eliminated to prevent them harbouring contamination*
- *Can we better protect areas that need to be kept clear of swarf/dust/oil, etc.*
    - *simple covers or enclosures*
    - *deflector plates*
    - *'T' slot covers*
    - *plastic caps*

Fig. 4 Set-up reduction: cleandown analysis guide

Adjustments (A): Aim to eliminate adjustments other than those which are fixed for the variety of components being produced. Off-the-shelf machine tools tend to have infinitely adjustable positioning methods requiring operations to measure for new positions and then possibly try a 'first off' with scrap and rework risks – go for deadstops/dowels, etc. where the one movement to load also automatically positions.

Problems (P): Need here is to get to the *root cause* and eliminate it so that it never re-occurs. (This approach and some simple analysis methods were taught during the training programme.)

At this first team meeting be ready for a mass of good ideas. Cummins found that they 'flowed like water' – but don't conclude the meeting until all of the ideas are classified into a prioritised action plan, added to the analysis sheet, and responsibilities and timing agreed. The aim is to get all of those involved at the meeting to volunteer to accept at least one assignment and not to simply generate a 'shopping list' for others.

Meetings continue, perhaps one per week, while improvements are installed and tested, until the team believes the method has been improved (and the duration of set-ups reduced in compliance with the goals set) sufficiently to demand that another video be made of the new method. This video is important as it serves a number of vital needs, for example:

- It gives recognition to the success of the team's activities.
- It serves as an invaluable training aid – both to new improvement teams and to new operators of the subject machine or equipment.
- It is the 'springboard' for further improvements. (Ed Hay's lengthy experience suggests that if the new method is allowed to mature and become the 'norm' for several months, then further improvements of significant magnitude can be made by a new or reconvened team.)
- It constitutes the basis for 'standardising' the new method via a concept called 'factor control'.

Standardising the improvement

As already mentioned, the team's brief is not only to reach an improvement goal, but to ensure that the method becomes the standard method of working (until further formally improved). *Factor control* was chosen as the vehicle for meeting this need.

Basically this is an enhanced form of process documentation, the main difference between it and the type of documentation normally used to detail a process in manufacturing is that factor control, utilising the skills and knowledge of the team, identifies the *critical factors* governing the integrity of the operation, and the parameters within which each factor needs to be controlled to maintain that integrity. An

SET UP AND CHANGEOVER FACTOR CONTROL SHEET		WORK TEAM BLOCKS TEAM A	FROM COMPT. (DESC) KV12 CYL. BLOCK NUMBER 3012345		
		WRITTEN BY C. BROWN	TO COMPT. (DESC) KV16 CYL. BLOCK NUMBER 3012346		
		DATE 25.1.86, REVISION N° -	MACHINE (DESC) ROUGH MILL PLANT N° 1234		
STEP NUMBER	WORK PERFORMED	CRITICAL FACTOR	PREVENTATIVE INSTRUCTIONS	LIMITS AND / OR FREQUENCY	CORRECTIVE INSTRUCTIONS
10	USE 12" BRUSH TO REMOVE EXCESS SWARF FROM 6 DEFLECTOR PLATES	SWARF AROUND CLAMPS	BRUSH OUTWARD AWAY FROM HINGE TO AVOID BRUSHING SWARF ONTO CLAMPS	EACH DEFLECTOR PLATE EVERY SET-UP	REWORK ANY INGRESS OF SWARF FROM CLAMPS USING VACUUM CLEANER WITH 90° NOZZLE
20	ACTUATE 6 OFF CAM ACTION CLAMPS TO RELEASE	CLAMP ANVIL CLEAR OF PAN RAIL LEDGE	ENSURE FULL RETRACTION OF CLAMP (HANDLE VERTICAL)	EACH CLAMP	VISUAL CHECK FOR ALL 6 CLAMP HANDLES TO BE IN VERTICAL POSITION
50	ACTUATE HYDRAULIC PUSHER CONTROL BUTTON	SYSTEM PRESSURE TO ENSURE FULL STROKE ACHIEVED	CHECK GAUGE PRESSURE READING	30-35 PSI EACH SET-UP	1. LOW PRESSURE: ENSURE HYDRAULIC FLUID IS AT CORRECT LEVEL (SIGHT GAUGE). TOP UP IF NECESSARY 2. ANY OTHER PROBLEM - REPORT TO MAINT. AND SUPERVISOR

SUPERVISOR _____   PROCESS ENG° _____   QUALITY AUDITOR _____

Fig. 5 Set-up and changeover factor control sheet

example is perhaps a hydraulic system (of clamping, say) where clearly the system pressure would be critical. The factor control concept requires that the apparatus should be equipped with a means of validating the pressure which is then required to be checked for compliance to the specification (say 30–35psi) at a specified frequency. All of this information where relevant is documented, as are the basic *corrective* actions required of the operator if the specifications cannot be met (see example, Fig. 5).

Status of the SUR programme

A number of teams are now operating within the Daventry Plant, and several have been formally 'wound-up' having, in the majority of cases, gained results better than their original targets (see summary of improvements and costs, Table 1).

The very first team to be formed had, within weeks of its inception, reduced the set-up duration on a £500,000 machine which drills and reams the cylinder head bolt holes on the cylinder block from over 17 minutes to just *8 seconds* for an outlay of £82 (Fig. 6). As an added bonus (and unforeseen at the outset) the work done by this team on their 'own' machine *totally* eliminated set-up on five other machines, and partially reduced set-up on a further five machines elsewhere in the production line. This was attained by a small alteration to the product, which itself cost nothing to install, but yielded a direct labour reduction of several thousand pounds annually.

Table 1 Machine set-up reduction achievements

| Machine | Set-up duration (mins) | | Reduction (%) | Expense (£) |
	Before	Now		
Head face drill	17.32	0.13	99	82
KTM rough borer	22.38	5.50	76	95
Ingersoll mill borer	15.47	2.18	86	35
KTM 2500/CINTI 10HC*	19.67	0	100	20
NC Webster & Bennett	159.72	27.00	83	930
Wavis drill (BRG CAPS)	4.72	1.52	68	0

*Projected

This kind of success, so early in the process, generated enthusiasm and competition in subsequent teams – so much so that informal teams began to emerge spontaneously and start work ahead of the formal training schedule.

The achievement of the teams on the cylinder block line specifically, enabled batch sizes of the 12-cylinder and 16-cylinder product to be reduced to less than one third of their previous size (from 20–25 to around the 6–7 level, dependent on product mix demand). This has led to shortened lead times, increased responsiveness, flexibility to meet short lead time mix changes and reductions to expensive inventory holdings.

Fig. 6 View of the head-face drilling machine

The drive to reduce set-ups continues in the miscellaneous product machining areas, and will be accelerated further in the near future as product lines are rationalised around new machinery additions and other changes in pursuit of a true customer-led Just-in-time manufacturing environment. No-one now sees set-up as an insurmountable barrier in pursuit of JIT – we have proven it can be reduced to an insignificant level anywhere it occurs, quickly, and at very low cost!

GROUP TECHNOLOGY AND JIT

M.P. Karle
University of the Witwatersrand, Republic of South Africa

The implementation of group technology can result in substantial benefits over the more traditional manufacturing systems, such as the job shop. With the Western world recently devoting much attention to the Japanese Just-in-time manufacturing philosophy, possibly the greatest advantage of group technology is now beginning to be acknowledged, i.e. a group technology manufacturing system forms the foundation stone upon which a successful JIT philosophy is built. This paper details the implementation of group technology at the Afrox Gas Equipment Factory in Germiston, South Africa, and how this system became the base upon which the newly adopted JIT philosophy is being built.

In December 1962, African Oxygen Ltd (Afrox) opened the Gas Equipment Factory (GEF) in Germiston, South Africa. The GEF was to manufacture gas welding equipment which had, until then, been imported from the parent company, British Oxygen (BOC). Numerous variations of the following four products were to be manufactured (Fig. 1):

- Gas pressure regulators.
- Gas cutting/welding torches.
- Gas cutting/welding nozzles.
- High- and low-pressure gas cylinder valves.

From its inception, the GEF was laid out as a traditional job, i.e. machines were grouped according to function. By 1982, the factory had grown from a 2400m^2 to a 9500m^2 plant and the total number of employees had risen from approximately 80 to over 400, the direct to indirect employee ratio being 3.7:1. By this time, the GEF had captured the major share of the South African market, the balance being made up by imported equipment. Customer demand for special-purpose products resulted in over 1000 variations of the four end-products being manufactured on some 300 automatic and manual machine tools. Inventory holding gradually grew to a total value of

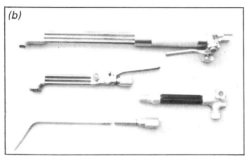

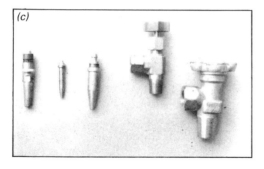

Fig. 1 Products manufactured at GEF: (a) gas pressure regulators, (b) gas cutting/welding torches, and (c) gas cutting/welding nozzles and high- and low-pressure gas cylinder valves

R3.2 million, thereby allowing the factory to operate with six months' inventory cover.

With the emergence, in the early 1970s, of high-quality, cheap products from Europe and above all, the Far East, pressure was brought to bear on Afrox by BOC to reduce costs at the GEF. In 1979, a MAPICS material requirements planning (MRP) system was installed in the GEF. The system was gradually upgraded with the implementation of more software packages, so that by 1984 the MRP system had grown into a successful manufacturing resources planning (MRPII) system.

Due to the initial failure of the MRP system, Afrox Management was forced to consider alternative methods of cost reduction. The avilability of group technology knowledge and experience within the BOC group resulted, at the end of 1981, in the decision to implement group technology (GT) as a means of reducing manufacturing costs.

Initial objectives of implementing GT

The main objective of implementing GT at the GEF, was to reduce manufacturing costs and thereby increase the return on investment (ROI). This could be achieved by being successful in a number of other objectives which, together, would result in lower costs and improved ROI.

By June 1982, a list of objectives was formulated. Some of these objectives were to:

- Centralise the responsibility and accountability for quality, tooling, completion by due date, etc.
- Rationalise the number of saleable products from over 1000 down to 300 and eventually 200.
- Standardise the use of purchased and manufactured components.
- Eliminate, as far as possible, the need for second machining operations on all components.
- Reduce machine set-up time to a maximum of 10 minutes for similar components.
- Shorten manufacturing lead-times and improve manufacturing flexibility and customer service levels.
- Train employees in new concepts and enrich their jobs by improving their skills and knowledge.
- Create an environment in which workers can identify with the end-products.
- Improve product quality and simplify quality control conducted by the machine operators.
- Improve machine maintenance and ensure maximum machine availability.
- Simplify production and inventory control and create a system of steady-state production with simple visual controls.
- Reduce total inventory holding, i.e. raw materials, work-in-progress (WIP) and finished goods inventory, to less than one month's cover.
- Improve floor space utilisation.
- Reduce material handling and the need for special material handling equipment.

Implementation and results of GT

The first step in implementing GT, was to determine the product families and therefore the GT cells. This was done by visual inspection, resulting in the establishment of four cells:

- Regulator cell, for the manufacture of all gas pressure regulators.
- Torch cell, for the manufacture of gas cutting/welding torches.
- Valve and nozzle cell, for the manufacture of high- and low-pressure gas cylinder valves and gas cutting/welding nozzles. The valves and nozzles were grouped together since, at the GEF, nozzle manufac-

ture is a flow-line manufacturing process which could easily be accommodated in the smaller valve cell.
* Central workshops cell, for the manufacture of special-purpose products on a quotation basis, while at the same time serving as a training centre for machine-tool setters and operators. Due to its unique nature, this cell would operate separately from the rest of the factory and would have a different reporting structure.

During December 1982, the first phase in the reorganisation of the factory was undertaken. Over the traditional Christmas holiday period, all heavy machinery was relocated in the cells. The factory operated in this state until the Easter weekend in April 1983, when during the four-day break, all the remaining machinery was relocated in the cells.

Each cell was established as an independent profit (not cost) centre and, as such, 'four factories within a factory' were established. A cell manager was appointed for each cell, who was accountable for all cell inventory holding, quality, tooling and completion by due date.

Within the four large cells, numerous smaller cells, or mini-cells, were incorporated, typically consisting of five machine tools run by one or two operators. These mini-cells manufacture a small family of components and result not only in reduced WIP and manufacturing lead-times, but also in accountability for output and component quality being centralised to one small area (Fig. 2).

The central material store was disbanded and all materials moved into the cells. Each cell has a component and a raw material store which is run by a cell storeman. These storemen are made accountable for all inventory in their stores, and report directly to their respective cell managers (Fig. 3).

Fig. 2 Mini-cells typically consist of five machine tools run by one or two operators

Fig. 3 Each cell has a component and a raw material store which is run by a cell storeman

The central tooling and gauging store was broken up and all tooling was moved into the cells. All the tools and gauges which are required to manufacture a family of components are stored in lockable glass cabinets – the machine-tool setters being held accountable for all tooling in their cells. This change resulted in the purchase of a significant amount of additional tooling (Fig. 4).

Fig. 4 Tool cabinet

Table 1   Improvements at the Gas Equipment Factory from 1982 to June 1986

• Inventory holding (real terms)	70% reduction
• Product proliferation (range)	70% reduction
• Manufacturing lead-time (main-line items)	80% reduction
• Number of employees	50% reduction
• Floor space requirements	60% reduction
• Product quality (average)	50% improvement

The Quality Control department was dissolved and the operators, not roving inspectors, made accountable for product quality. Many low-cost jigs were manufactured for simplifying machining and assembly operations, thereby reducing lead-times and improving product quality.

In order to make the cells visually autonomous, in December 1983, each machine in a cell was painted the 'cell colour':

Brown: Regulator cell
Blue: Torch cell
Green: Valve and nozzle cell
Grey: Central workshop cell

By June 1984, each cell employee was issued with an overall which was either brown, blue, green or grey, depending on which cell he worked. Workers could now visually identify themselves with their cell and this, it was hoped, would enhance worker pride and improve industrial relations.

Since its inception in April 1983, GT has resulted in numerous quantitative and qualitative improvements at the GEF, some of which are listed in Table 1. It should, however, be noted that some of the reductions are not only due to GT, but also to the present economic climate and the recent success of the MRPII system.

Although difficult to prove quantitatively, the GT concept has resulted in improvements in worker pride, team spirit, job enrichment and greater worker satisfaction according to discussions with cell managers and other employees.

JIT development at the GEF

Since the implementation of the cells at the GEF in April 1983, substantial savings and improvements have been achieved. However, the system is by no means perfect and further improvements are always possible.

The highly successful Japanese JIT philosophy is ideally suited to the GEF since much of the groundwork for JIT has already been completed. Many JIT principles[1,2] are in fact GT principles[3] so it is quite fair to say that since the implementation of GT in 1982/83, the GEF has unknowingly been moving down the JIT road. The following are some of the common objectives of both GT and JIT:

- Reducing set-up time by grouping components in families and creating machining cells.
- Rationalisation and standardisation of products and materials.
- Reduction of all inventory holding, especially WIP inventory.
- Reduction of lead-times, e.g. by low-cost automation and WIP reduction.
- Increasing employee responsibility and accountability and thereby improving job satisfaction and product quality.
- Simplification of production, inventory and quality control by using simple visual controls.
- Ensuring maximum machine availability through preventive maintenance.

Early in 1986, JIT formally became the manufacturing philosophy of the GEF. The adoption of a JIT philosophy would build upon the success of the GT system and further reduce manufacturing costs. By February 1986, detailed training of management, supervisors and machine operators had begun.

In May 1986, the first GEF 'pull' system was implemented in the valve and nozzle cell. The high-volume, nozzle production line was selected since it is a flow-line manufacturing process and set-up time from one nozzle size to another is negligible. After extensive operator training, the line was started with a lot size of 100 nozzles which were 'pulled' through the 11 machines by means of kanban squares which had been painted on the factory floor. Within weeks the lot size had been reduced to the present 10, resulting in the benefits outlined in Table 2.

The results of this 'pull' system far surpassed expectations and resulted in the rapid development of more such systems. This required the relocation of certain machine tools in order to set up more mini-cells, which could then be operated on a 'pull' system.

Building on the early success, further JIT-related ideas are being implemented. These include greater vendor involvement in raw material supplies and greater employee involvement / communication with regard to various aspects of manufacturing, especially quality

Table 2 Results of the 'pull' system for nozzle production

• Work-in-progress inventory	80% reduction
• Manufacturing lead-time	90% reduction
• Output	40% increase with no overtime and four operators less
• Scrap	70% reduction
• Resource utilisation	Significant improvement
• Employee contributions towards safety, production, and quality problems	Significant improvement

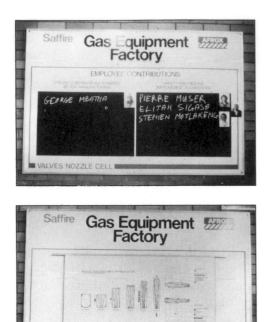

Fig. 5 Example 'measles' charts

improvement. The implementation of large 'measles' charts (Fig. 5) had a significant effect on quality – in one particular case scrap dropped by about 50% in one week. The adoption of a successful JIT philosophy has now become the major objective of the GEF.

Concluding remarks

Group technology is often regarded as a prerequisite for a Just-in-time philosophy[1]. The implementation of GT at the GEF in 1982/83 resulted in tremendous quantitative benefits and also created a factory which could readily adopt a JIT philosophy. The JIT philosophy can now build upon the success of the GT system and in so doing reduce manufacturing costs and increase the chances of economic survival. Results at the GEF have shown that Just-in-time is dependent upon a successful group technology system.

References

[1] Schonberger, R.J. 1982. *Japanese Manufacturing Techniques, Nine Hidden Lessons in Simplicity*. The Free Press Division of MacMillan, New York.
[2] Hall, R.W. 1983. *Zero Inventories*. Dow-Jones Irwin, Homewood, Illinois.
[3] Burbidge, J.L. 1975. *The Introduction of Group Technology*. Heinemann, London.

3

PRODUCTION/MATERIAL CONTROL

The papers included in this section describe the various techniques and approaches used in adapting production/materials control to support JIT manufacturing. Three areas are covered: production planning and control, production smoothing and flow analysis. The papers on production planning control look first at how traditional materials requirement planning (MRP) systems can be adapted to a JIT environment. This is followed by a new approach to planning and control: OPT. The next paper looks at how production can be smoothed and how this can be measured. The final paper, looks at the technique of flow analysis.

DEVELOPING AND IMPLEMENTING CONTROL SYSTEMS FOR REPETITIVE MANUFACTURING

M. Sepehri
University of Southern California, USA
and
N. Raffish
Deloitte Haskins & Sells, USA

Repetitive manufacturing is the fabrication, machining, assembly and testing of standardised units produced in volume, or of products assembled to order in volume from standard options. It is distinct from production in the process industries in that it deals primarily with the handling of discrete units, and less with fluids, powders or processes involving chemical change. It differs from job-shop manufacturing primarily in the volume and sequence of units produced and in having a preset production rate.

To contrast repetitive manufacturing with other types of production, Table 1 presents a framework of the different types of manufacturing as distinguished by physical characteristics and methods of control.

There are many similarities in the way production is planned for all types of manufacturing. Every type involves production planning; master scheduling and capacity planning are often involved also. The differences show up in the way production is controlled, and sometimes in the way the controls are organised.

Table 1 Examples of different types of manufacturing

Manufacturing type	Product industries (discrete units) Fabrication and assembly	Process industries (non-discrete units) Processing or unit operations
Project manufacturing	Construction Shipbuilding	Specialty chemicals
Job-shop manufacturing	Machine tools	Pharmaceuticals
Flow manufacturing	RV receivers Box wrenches (repetitive)	Fertiliser (continuous)

Over 20 electronics companies in the Silicon Valley of California[1] were studied to identify the general characteristics and requirements of repetitive manufacturing. Although all the companies classify themselves as repetitive manufacturers or interested in utilising repetitive systems, there are a number of differences in their conditions, priorities and degrees of correspondence to the definition of repetitive manufacturing.

Common characteristics

The repetitive manufacturing environment is typically characterised as follows:

• Production runs are long, continuous and high volume, rather than made up of discrete lots.
• Production lines tend to be fixed; machines and people are grouped according to the families of product produced, and item flow is along a fixed route.
• Standardised products are produced, and materials are often issued to the line in production or bulk quantities rather than in 'kits' or discrete lots.
• Blanket schedules are used instead of specific manufacturing orders, with emphasis on volumes produced over time.
• Products are commonly processed 'in-line'; fabrication and subassembly are often performed simultaneously with the final assembly operation.

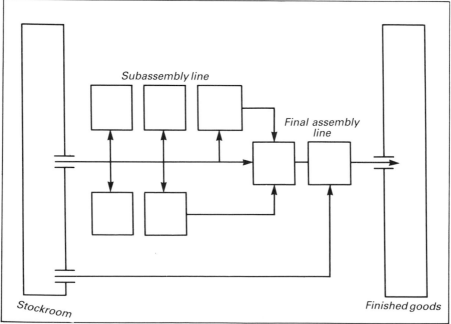

Fig. 1 Layout of repetitive production line

• Operations are often combined in single work centres.
• In a steadily flowing process, it is difficult to associate lead-times with products being produced.
• Operations tend to be capital intense, and special purpose equipment is often used.
• Work centres generally provide highly specialised operations which are portions of the total build-up sequence.
• Fixed costs are relatively high; however, unit direct labour and sometimes material costs are low.

Generally, a production line is set up for a product or product family. Capacity and tooling are dedicated to individual work centres, thus eliminating competition for resources among various products or product families. In addition, routings are usually fixed, and operation layouts are designed for minimal material movement and transfer.

Fig. 1 illustrates the flow-through nature of the manufacturing process in a typical repetitive manufacturing production line. In a production line in which several work centres are linked together, the production rate is established for optimal overall production.

In discrete manufacturing, the manufacturing process is authorised by the work order. It includes a shop paper that describes the routings of the assemblies to be made, the relevant bills of materials (BOMs), engineering drawings, and specific tool and fixture information.

The work order is eliminated in repetitive manufacturing, because it is not necessary to maintain quality integrity at a detailed level. The flow order replaces the work order as a method of scheduling quantities of components.

A flow order is an order or requirement from material requirements planning (MRP) that is filled over time and checked by a cumulative count until the assembly requirement is complete. The flow order provides the master production schedule (MPS) and MRP with the replenishments needed to meet the desired production rate over a certain period, which may range from a day to a month or more.

Problems with work-order-based systems

Due to the particular conditions of repetitive manufacturing outlined earlier, MRPII must be modified for the repetitive manufacturing environment to handle the highly specialised problems created in such a system. Some of these problems are that:

• Material lot integrity is not always maintainable.
• Maintaining shop-floor reporting is difficult.
• Detailed job status reporting is not possible.
• Performance must be monitored without depending on discrete manufacturing order mechanisms.
• Work performed (but not completed) on one schedule is often transferred to updated schedules.

- Inventory record balances are often less accurate than those kept by discrete manufacturers.
- Schedule changes may be frequent and very rapid.

Other problems with using work-order-based MRP systems in a repetitive environment are described below.

Scheduling system interfaces

In a discrete manufacturing environment, the MPS is used to correlate final assembly work orders with finished goods lots. With flow order control, however, the scheduled receipts concept, which is used in MPS both to replenish inventory to meet customer demand (or forecast demand) and to establish requirements for lower level demand, is inappropriate. In addition, fixed routings and dedicated tooling and capacity in repetitive manufacturing create a system in which each order is processed independently. Repetitive manufacturers, therefore, manage capacity for specific tasks or products and have little practical use for modules such as shop-floor control.

Product costing

The process of determining true product cost in discrete manufacturing is inexact because the cost of assembling an item varies. If items are made infrequently, the start-up learning curve will vary. If jobs are set aside for a priority job, their cost will be different from the cost of processing them directly.

The same is true if machines with different speeds are used at various times to produce an item. If the variations are small, they may average out, but if they are large they can cause accounting problems. If the true cost is unknown, management's profit objective will not be met. Inaccuracies can lead to a selling price that is too high, and perhaps not competitive, or to one that is too low and therefore unprofitable.

Product costing in a repetitive shop is more complex than that in a discrete shop – not only because transactions that affect direct product costs cannot be associated with a work order, but also because of product sequencing. Repetitive manufacturers employ product sequencing to produce similar products or product families. Repetitive shops usually produce several items on the same product line set-up. Because the different models are not identified by separate work orders, direct product costing is impossible.

Production reporting

In a discrete environment, the work order is updated each time an operation or a job step is completed, when a split-or-move lot occurs, and when the work order is closed. Labour charges on the shop-floor are recorded on the work order to account for product cost. These labour charges, in addition to all direct costs, are charged to the work order and relieved from WIP.

In repetitive manufacturing, however, a flow order has daily completions and might be open for days, weeks or months. Because certain costs are unknown when completions are reported from production, WIP is relieved only on a standard or estimated cost basis. In addition, material usage reporting is imprecise when repetitive manufacturers use MRP.

Material planning and control

Since discrete manufacturing systems schedule material arrivals to coincide with the start of the work order, work order control typically creates a need for extensive expediting to ensure efficient plant operation. Staging material at the work centre creates queues and excessive WIP inventory that often must be expedited to maintain established production priorities. This process is very costly.

Expediting is inappropriate in repetitive manufacturing, because a major objective of repetitive manufacturers is to minimise WIP and raw material inventories. For repetitive manufacturers to gain the full benefits of MRP, most – if not all – of the material planning, ordering, controlling and issuing of subsystems must be enhanced, both to support a continuous (flow order) material consumption environment and to accommodate just-in-time vendor delivery scheduling.

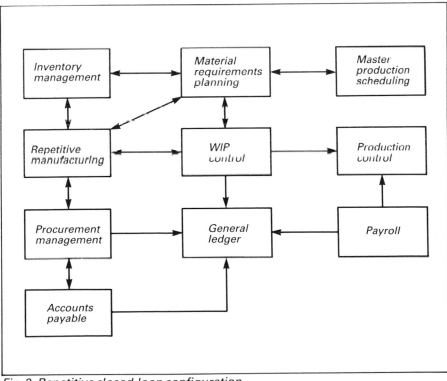

Fig. 2 Repetitive closed-loop configuration

Computer systems development

Repetitive manufacturers who want to obtain the full benefits of a work-order-based manufacturing control system such as MRP must implement a repetitive manufacturing software module. Repetitive software must be integrated into the full control system and should meet the specific needs of the repetitive manufacturer. This software therefore becomes one of several modules in a closed-loop integrated manufacturing control system (Fig. 2).

Integrating repetitive manufacturing software into the system requires several key linkages:

- Inventory control software must be enhanced to record WIP record balances accurately and indicate workcentre–part relationships.
- MRP and master production scheduling software must recognise flow orders and support the allocation functions used by repetitive manufacturers.
- Product cost records may need new data elements for workcentre overhead rates, depending on the stability of the product line and the production rate, to account for variable overload rates so that costing will be accurate.
- Procurement software may need to be enhanced to receive a purchase order directly at a workcentre. In addition, automatic requisition release from MRP bases at near-term flow rates may be desirable.

The repetitive manufacturing function must be integrated into the manufacturing control system. Because transactions must be recorded on a timely basis to keep the WIP balances and allocation records current, a separate or redundant inventory database would be counterproductive.

A defined set of procedures for repetitive manufacturing planning and control is still evolving, and an extensive software development and experimentation period will continue for several years. There are, however, certain required features and operations that should be supported by an effective repetitive software module. The following sections describe the software features and operations repetitive manufacturers need.

Workcentres

The workcentre is the point at which WIP flow is monitored and measured. It can be a machine, a set of linked processes or an inspection operation. Two or more linked workcentres constitute a flow line. The workcentre records transactions and completions and serves as a gate through which production flows. In addition, the workcentre stores such data as lot control information and labour transactions.

The system should also provide a data element for maintaining WIP balances. This feature monitors and controls material allocations, issues and completions by workcentre and enables accurate cycle counting by workcentre.

Flow order

The flow order establishes the requirements for the master production scheduling and MRP systems and also records a workcentre's completions on an assembly. To balance production requirements at each workcentre, flow orders are also created for each subassembly to be manufactured. Material allocations are established and distributed at lower levels to fulfill the flow order requirements. Because of the integration of the flow process, material is often allocated to cover several workcentres at one time.

The automatic multilevel allocation and flow order creation process should be implemented gradually to allow time for adapting organisational procedures. When the production line is balanced and a daily production rate is established, however, these techniques will relieve the repetitive manufacturing planner of much of the tedium of manual calculations.

Material routing

The use of repetitive manufacturing MRP complicates product costing, because there is no work order to track direct costs. After completing a certain quantity and variety of assemblies, a repetitive manufacturer calculates the materials consumed based on the bill of materials and estimates labour charges using labour standards. Material routing, a combination of the BOM and operation routing, is an integral part of inventory control system software (Fig 3)

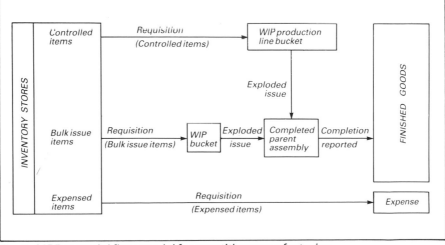

Fig. 3 MRP material flow model for repetitive manufacturing

Including the routing on the BOM enables planning and production departments to deliver material or locate material already in the production process. If repetitive manufacturers can easily locate material and components, costs can be assessed for each assembly build-up, and the value of partial assemblies can be calculated. This also permits accurate costing for scrapping of partially built assemblies.

Backflushing

Although transaction recording is necessary for accurate product costing, workcentre assembly completions often occur so quickly that it is inefficient and unnecessary to record each one. Depending on the assembly build-up, it may be sufficient to report a completion after every fifth workcentre, once an hour or once a day, for example. When labour has been performed and material added at the intervening workcentres, backflushing is used to draw forward the proper components and cost from the last reporting workcentre.

In backflushing, the system accesses the BOM and standard cost databases and combines the costs of the components that have been used in the assembly build-up since the last reported completion. Fig. 4 illustrates this procedure. Depending on the complexity of the assembly, the system can be programmed to report only the completion of the finished parent assembly.

In most repetitive environments, backflushing occurs at several points along the production line so that plant management has access to current production status. The number of reporting workcentres is related to the flow-through time of the assembly build-up. For example, the faster an assembly moves through the shop, the less important it becomes to report each movement or interim production status.

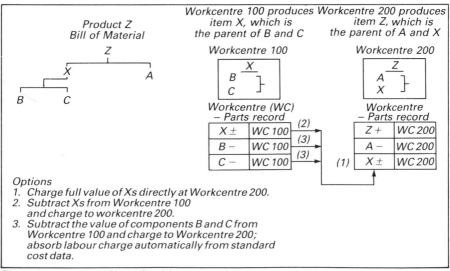

Fig. 4 Illustration of backflushing procedure

Time-phased allocation and usage

The integration of time-phased material allocations into the MRP software system allows repetitive manufacturers to reduce WIP inventory and queues, lead-times, and the amount of time and money devoted to expediting activities. These improvements increase management's ability to monitor and control shop-floor activity. For example, if a workcentre processes 900 components daily but has a floor constraint (station capacity) of 300 units, the allocation system divides the requirements, indicating that three daily line deliveries of 300 units each are needed.

If a component is used at more than one workcentre, MRP continues the aggregate planning; the allocation process segments the aggregate planned orders into the quantities needed to support multiple flow rates. In planning the size and timing of deliveries, factors such as cost, storage space and dock-to-stock flow-through rates must be weighed, considering the manufacturer's budget, storage facilities and operations.

As the maturity of the users increases, material issues will become controlled more by completion (pull) than by designated orders (push). This will provide tight control over WIP inventories.

Implementing the system

The most difficult part of implementation is changing the informal work-order-oriented system that employees have developed gradually to operate the formal system. The potential impact on the physical plant is significant, and the training and educational and organisational changes required to introduce repetitive software into the system must not be minimised.

One organisation that recently converted part of its product lines to repetitive software spent one year developing requirements and another year implementing the necessary organisational changes. A high-technology manufacturer devoted almost two years to experimenting with and refining processes and educating employees. As an example, in the early stages of implementation, the level of detail of workcentre activity will vary greatly until the users discover what data are needed.

In addition, vendor-user relationships need to be changed for successful repetitive software implementation. Longer term projections to vendors must be established – for example, by providing the vendor with a 90-day purchase order report and a nine-month requirements forecast. In addition, engineering, material and operations functions must develop a uniform strategy to support the allocation process.

If the repetitive manufacturer lacks an integrated database, it will be difficult to create major enhancements to support repetitive software.

A basic set of user-oriented report generation tools is useful in the early stages of implementation. Such tools alleviate the demand on manufacturing information systems for ad hoc reporting by users. If each user can extract the data he or she needs, the MIS department will be more efficient.

Justifying the implementation of a repetitive manufacturing software system is both simple and difficult. It is easy to determine the development, education and installation costs and weigh them against expected reductions in the various inventories.

The difficult part of determining payback involves analysing the intangible benefits. These benefits can be derived by asking questions such as: Can the current system meet production needs? Are the machine start-up and labour efficiency factors always unpredictable because of the load and mix on the floor? Can the current way of building products be made more productive? What is the competition doing? Senior management must address these difficult questions.

Although repetitive manufacturing software has been implemented in only a few manufacturing organisations, it has potential uses for both repetitive and discrete manufacturers. With the application of such techniques as group technology (GT) and Just-in-time inventory control, in addition to automated manufacturing processes, production runs gradually will become shorter and more efficient for discrete manufacturers. Therefore, although the repetitive manufacturer will gain immediate benefits from repetitive software, those benefits will filter down to discrete manufacturers as production becomes more standardised and efficient.

Just-in-time

It is technically possible to install a repetitive system without any concern for Just-in-time (JIT). If this is done, the company is likely to wind up with as much WIP inventory and as many record keeping transactions as it had originally.

The reason for this is that the disciplines needed to implement and maintain a repetitive system are very closely allied to JIT. Balancing the workcentres really requires that only enough inventory to sustain the flow rate for that line be available on the shop-floor. The completion rates are predicated on little or no rework activity, which would disrupt the flow and the flow rate. This means that in a repetitive system, quality becomes extremely important to meeting the designated flow or completion rates.

An element of this consideration is vendor quality. In fact, vendor quality and on-time performance are critical in a repetitive environment. Since one of the objectives of a repetitive system is decreasing WIP and stores inventory, the risk of having production interrupted by late vendor deliveries is significant.

Any discussion of JIT generally centres around the broad definition of waste elimination. In a repetitive system in which one of the goals is to be able to produce smaller quantities with little or no cost impact, items such as engineering configuration changes, set-up process changes and multifunction worker training will bear heavily on whether a company can increase flexibility and hold costs level. Poorly timed material or process changes can cause waste not only in terms of output rates, but in terms of scrap and rework as well.

A repetitive manufacturing system that has incorporated the JIT philosophy will be able to reach its payback goals within a reasonable timeframe. A repetitive system that has not will gain a reputation as a system that has great promise, but was too difficult or too cumbersome to operate, and therefore the objectives were never attained. However in retrospect, an objective observer might attribute the lack of success to a lack of discipline and of the commitment to sustain that discipline in a 'no waste environment'.

Reference

[1] Sepehri, M. 1983. Integration of Western and Japanese management techniques in repetitive manufacturing. In, *Proc. American Decision Sciences Conf.*, Hawaii.

OPT

LEAPFROGGING THE JAPANESE

R.E. Fox
Avraham Y. Goldratt Institute, USA

OPT has been used to categorise a set of logical concepts for running a manufacturing business and describes a software system for scheduling and simulation. The OPT approach is essentially a logical derivation of The Constraints Theory developed by Eliyahu M. Goldratt. This paper describes the concepts used by the OPT software system and how it executes the OPT concepts. Contrasts with MRP are drawn to illustrate similarities and differences. The OPT approach basically involves first identifying the contraints of the system, scheduling them finitely to fully exploit them, and then subordinating all other activities to these decisions.

OPT is a powerful new approach for managing manufacturing organisations. Its philosophy and tools can be applied in several ways to better manage our companies. This paper deals with the use of OPT as a scheduling and simulation tool, where it can be thought of from five different viewpoints:

- As a philosophy for scheduling.
- As a language for modelling manufacturing operations.
- As a software package for master scheduling, material and capacity requirements planning, and detailed scheduling.
- As a finite scheduling module within this software package which develops optimised production schedules.
- As a way of focusing data accuracy efforts to obtain significant benefits before reaching high levels of data accuracy.

The OPT system follows the nine principles of good scheduling shown in Fig. 1. In addition, there are three criteria that schedules must meet:

- Move the organisation toward the goal of manufacturing – making money.
- Be realistic.
- Be immune to disruption.

1. Plant capacity should not be balanced
2. The utilisation of a non-bottleneck resource is not determined by its own potential, but by some other constraint in the system
3. Activating a resource is not synonymous with utilising a resource
4. An hour lost at a bottleneck is an hour lost for the total system
5. An hour saved at a non-bottleneck is a mirage
6. Bottlenecks govern both throughput and inventories
7. The transfer batch may not and many times should not be equal to the process batch
8. The process batch size should be variable and not fixed
9. Schedules should be established by looking at all of the constraints simultaneously. Lead-times are the result of a schedule and cannot be predetermined

Fig. 1   OPT's nine principles

In order to meet these goals, all scheduling systems must follow these principles. These rules are not just the rules of OPT, but the true rules of nature. In earlier papers we have seen how these rules are followed by the Just-in-time system [1-4]. It can also be shown that other efficient methods of production such as assembly lines and process industries follow OPT's nine principles. The remainder of this paper will show how the OPT system works and utilises these rules.

Modelling a manufacturing plant

The first step is to construct a model of the manufacturing facility to be scheduled. The model is essentially a network of how the various manufacturing resources (people, machines, tools, etc.), customer orders, products and raw materials are linked together.

Fig. 2 represents an illustration of such a model or network. At the top of the diagram are the market requirements (either forecasts or firm orders) which are linked to the various assemblies or products manufactured. These products, in turn, are linked to the subassemblies from which they are asembled. The subassemblies are chained to the parts from which they are produced and the various manufacturing processes through which these parts must move. Finally, the first manufacturing operations are tied to the appropriate raw materials. The network can be created to whatever level of detail necessary. It can include interactions between machines, fixtures and gauges, as well as direct labour workers, set-up workers and inspectors.

Alternative operations and even alternative bills of materials can be realistically modelled using the language of OPT. In fact, only 24 data fields are necessary to fully describe any manufacturing operation in such a network. Changes to the model or network can be made readily so it always remains an accurate reflection of a particular manufacturing facility.

One of the powerful aspects of OPT is that the modelling language allows it to describe accurately an enormous array of manufacturing operations. As of this date, no manufacturing operations have been encountered which OPT has not been able to model successfully and accurately.

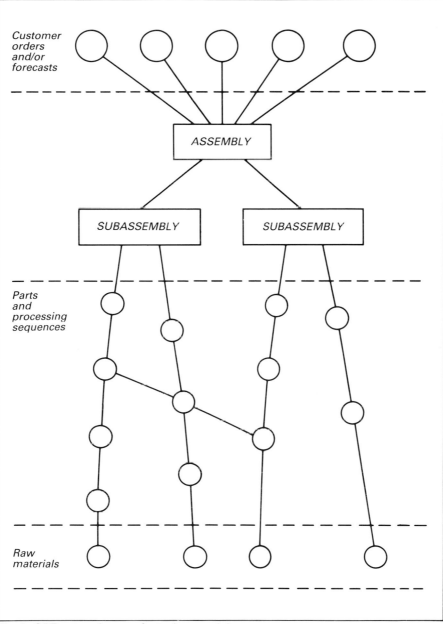

Fig. 2 OPT network manufacturing model

The creation of the initial model of the manufacturing facility (the OPT network) is usually accomplished using data generally available on a company's computer system. Existing data files containing bills of materials, routings, market requirements (sales forecasts and firm orders), inventories and workcentre data (set-up and processing times) is fed into a module of the OPT system called Buildnet. Buildnet takes this data and creates the OPT network. It essentially links the data together in a different fashion.

Buildnet also highlights logical errors in the data for correction, such as bills of materials that do not link to routings, inventories for which there is no part number, or customer requirements for non-existing products. These logical data errors are identified so the gaps can be filled and all connections made in the OPT network.

Finding the bottlenecks

Once this network or model of the manufacturing facility has been constructed, the next task is to determine where the constraints or bottlenecks are in the system. This is accomplished through an iterative process which begins by running this network through a module in OPT called Serve. Serve is OPT's enhanced version of an MRP system.

One of the outputs of the Serve run is a load profile for each of the resources in the model (Fig. 3). These load profiles are the typical patterns that result from traditional MRP systems. They look somewhat like the Manhattan skyline and have many peaks and troughs.

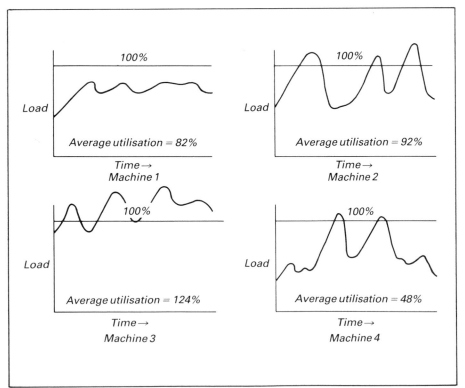

Fig. 3 Load profiles for the Serve module

The master scheduler is usually asked to level load by putting the peaks into the troughs. It is extremely difficult to develop a realistic master schedule that results in realistic schedules at all levels of the product structure. The best we can hope for is a 'doable' schedule – not an optimised one.

The Serve module approaches the problem from a different vantage point. In addition to generating the load profiles for each resource, it also calculates the average utilisation of each resource. These resources are then sequenced in order of their utilisation with the most heavily loaded resources ranked first.

At this stage, if the data were totally correct then the item at the top of the list would clearly be a bottleneck. Unfortunately, the data is usually bad. If we were to wait until all the data had a high degree of accuracy, we might have to wait for our grandchildren to take the next step. However, part of the OPT approach is to focus on the data that it is most important to have accurate. This is a very small percentage of the total data in a manufacturing company, frequently less than 1%.

OPT applies the ABC rule to data accuracy. The difficult task, however, is to determine which data are the A items. The OPT philosophy says the criterion we must apply is: "What impact does the inaccuracy of a data element have on our ability to make money, i.e. how does it affect the total system?" If it has a large impact then obviously it is an item whose accuracy we must establish now. If it has a very small impact on our ability to make money, then it can wait until more resources are available. Clearly, the bottleneck or a constraint in the system is a place where it is important for the data to be accurate.

In the example shown in Fig. 3, machine 3 appears to be the bottleneck. The first question we must ask ourselves is: "Does this machine exist, and do these parts actually go through this machine?" It is not uncommon to have out-of-date routings and to find when we go on the shop-floor that a particular machine has been scrapped or the parts are run on other machines. In these cases, what appears to be a bottleneck is merely a mirage. Let us assume, however, that the machine does exist and all the parts that comprise the load of 124% do go through that machine.

The next question we must ask ourselves is: "Can any of these parts be done on another machine?" If they can, then this alternative machine should be included in the network and that portion of the load removed from machine 3. Let us further assume that the parts cannot be off-loaded to another operation. The next step is to determine if the standards used for set-up and process times are correct. This is best done, not by looking at the engineering data or the cost standards, but by talking to the foreman and the workers that operate the machine. We frequently find in these circumstances that the set-up time is not fixed but variable. We may have dependent set-ups.

This situation can be illustrated through the analogy of a painting operation. When the paint operation is changed over from white to black, the set-up time may be a matter of minutes because the system does not have to be totally washed out. Any flecks of white paint left in the system will be covered over by the black. However, when we change over from black paint to white paint, it may take several hours to thoroughly clean the system and remove all black paint. If we do not, the black flecks will show up when we begin painting white parts.

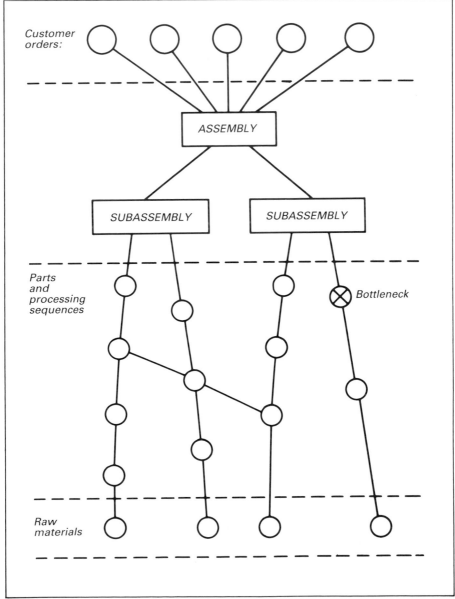

Fig. 4 OPT network with a bottleneck

When there are set-up dependencies, the logic and time required in these dependencies need to be built into the system to reflect the reality of what is happening. Finally, we need to check the accuracy of the bill of materials from this bottleneck up to and including the orders for the products made using this part (Fig. 4). Once these checks have been made, all the corrected data are reloaded and the OPT network updated. We should have at this time very accurate data on this

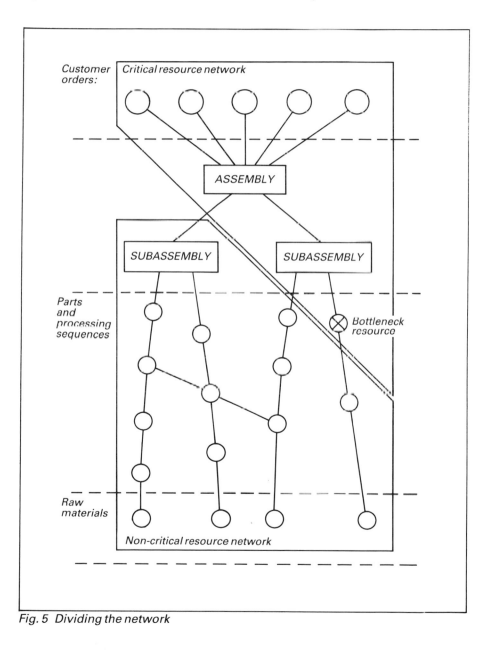

Fig. 5 Dividing the network

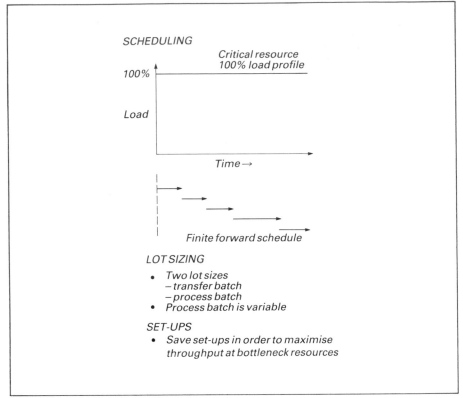

Fig. 6   The key features of 'Brain of OPT'

potential bottleneck operation and the other related areas which could impact the ability to make money.

At this stage, we will completely rerun the Serve system. Again, the average utilisation of each resource will be calculated and rank ordered from most heavily loaded to lightest loaded resource. If machine 3 remains at the top of the list, we know that we probably have isolated one of the constraints in the system. It is possible that, because of data inaccuracies, some less lightly loaded resource is in fact the bottleneck. If this is the case, it will surface at a later time. If machine 3 has dropped down lower on the list of resources then we will again take the resource that is most heavily loaded and repeat the above process. For the sake of simplicity, let us assume that machine 3 has remained at the top of the list.

Scheduling the plant

Once a bottleneck is identified, the network is divided into two parts – the critical resources and the non-critical resources. The critical resource portion of the network contains the bottleneck operations and all other portions of the network, up to and including customer orders as shown in Fig. 5. The remaining portion of the network we call the non-critical resources. This dividing of the network is done by another module in the OPT software called Split.

The critical resource portion of the network will be scheduled using a module called the 'Brain of OPT'. The non-critical portion of the network will be scheduled using Serve. The Brain of OPT contains Dr. Goldratt's mathematics and generates optimised schedules for the critical resource portion of the network. (A detailed explanation of the essential logic of this drum-buffer-rope approach can be found elsewhere [5,6]. It develops a finite forward schedule for the bottleneck resources and the other resources in this portion of the network.

The Brain of OPT not only schedules production but also determines the transfer and process batch sizes at each operation. It attempts to maximise the throughput at the bottleneck operation while at the same time maintaining a synchronous flow of parts to ensure the correct mix of parts is being produced. At the critical resources, the Brain of OPT loads these resources to 100% of the available capacity and schedules them in a finite forward fashion (Fig. 6).

The output of the Brain of OPT is then input into the Serve system along with the non-critical portion of the OPT network. Serve now operates as a 'smart' MRP system. MRP is now valid since we are scheduling only resources that have excess capacity. One of the failings of the MRP approach is that it assumes unlimited capacity but is applied to environments where bottlenecks exist. This omission destroys the accuracy and usefulness of the resulting MRP production schedules and purchasing requirements. However, in this case we have only resources with excess capacity and therefore the MRP scheduling logic works.

Serve, however, has many features not found in traditional MRP systems. For example, it recognises that a non-bottleneck operation has excess capacity or slack and schedules these resources to provide roughly equal utilisation over time. As a result, this resource always has a buffer of 'safety capacity' available in case our friend Murphy strikes.

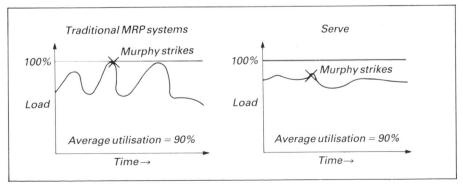

Fig. 7 Load profiles for MRP and Serve

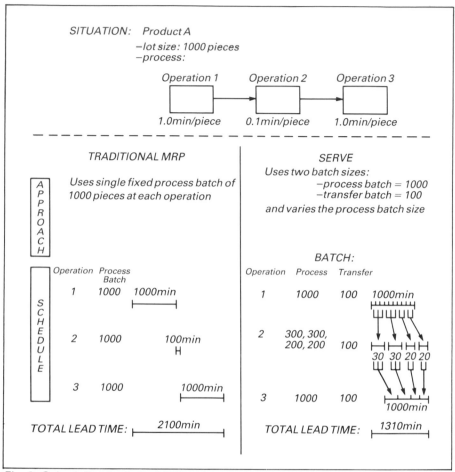

Fig. 8 Scheduling logic for MRP and serve

Traditional MRP systems generate schedules that look like a Manhattan skyline and do not equally divide up the slack capacity to provide a constant buffer. Unfortunately, our friend Murphy usually strikes at the peaks of the Manhattan skyline rather than the valleys (Fig. 7). It is at these peaks that he does the greatest damage and creates and propagates through our system, the famous floating bottleneck.

Serve also utilises transfer batches and process batches to generate overlapping schedules that significantly reduce lead-times. Fig. 8 contrasts how a traditional MRP system and Serve handle the same scheduling problem. The MRP system employs only a fixed process batch of 1000 pieces and takes 2100 minutes to process it through the three operations.

Serve employs two types of lot sizes – a fixed transfer batch and a variable process batch. At Operation 1, it uses a process batch of 1000 pieces but a transfer batch of 100 pieces. When each transfer batch of 100 pieces is completed, they are moved to Operation 2. In this example, Serve has scheduled process batches of 300, 300, 200 and 200 at Operation 2. Operation 2 is scheduled to work on its first batch of 300 pieces once the first three transfer batches have arrived. As Operation 2 completes each transfer batch of 100 pieces, it moves these parts to

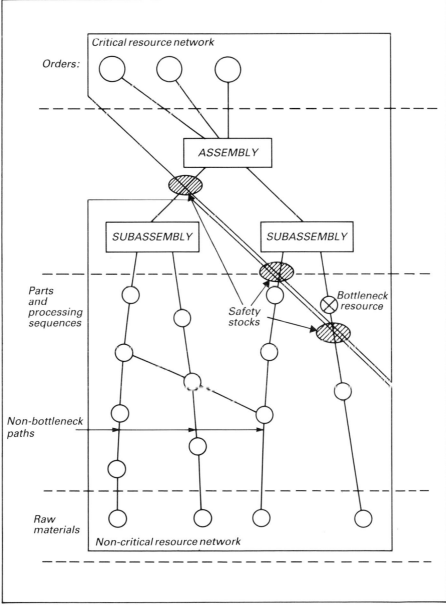

Fig. 9 Safety stock locations

Operation 3. As soon as the first process batch of 300 pieces is completed, Operation 2 is available to process other parts. Operation 2's second process batch of 300 pieces is scheduled to be started at 600 minutes after Operation 1 first started up.

The third process batch is scheduled to start at 800 minutes and the fourth and final one at 1000 minutes. Operation 3 begins its single process batch of 1000 pieces once the first transfer batch of 100 pieces arrives at time 310 and then works continuously for 1000 minutes. Additional transfer batches arrive in sufficient time to keep Operation 3 running smoothly.

The total lead-time to process the 1000 pieces through this scheduling approach is only 1310 minutes (a 38% reduction in lead-time) versus 2100 minutes for the conventional MRP approach. It should be noted that Serve is scheduling many more set-ups on Operation 3 than MRP – four versus one. However, as this is a non-bottleneck operation, there is no additional cost for these set-ups. When appropriate, Serve will schedule additional set-ups to reduce work-in-process inventory and lead-time. It will not schedule so many set-ups that a non-bottleneck operation becomes a bottleneck.

Protecting the schedule

Another important facet of OPT is how it locates safety stock throughout the system. The level of work-in-process in the system can be dictated through the use of a managerial parameter. However, OPT determines where that inventory should be best located in order to maximise throughput and ensure that the schedule is immune to disruption. As we have seen above, safety capacity is used as a buffer against Murphy. Inventory is also used as a safety stock against disruptions caused by Murphy. OPT schedules to place safety stock in two key areas (Fig. 9):

- In front of bottleneck operations.
- At the intersection of non-bottleneck paths and the path from a bottleneck up to its orders.

Jobs will be scheduled to arrive at these points some time in advance of when they are needed. The amount of time depends on the particular manufacturing environment, but from several days to a week is not uncommon. The purpose is to ensure that the throughput of the system is not impaired. Having work arrive at the bottleneck operation one week before it is needed will immune the bottleneck operation to any upstream disruptions that are less than one week. The same logic applies for work arriving at junctions to the path from the bottleneck up to the customer orders. This work should be scheduled to arrive in advance so delays will not disrupt throughput of the total system.

Once this process has been completed, if there are no other constraints or bottlenecks in the system, we have developed schedules that can be used on the shop-floor and meet the criteria for good scheduling. However, usually at the end of the first iteration, we find there are additional bottlenecks in the system. If so, these bottlenecks are identified because in the Serve schedules they exceed 100% utilisation. The data should be checked to ensure it is correct and if so that resource is moved over into the critical portion of the network and the above process repeated.

This process, which usually involves five or six iterations, is continued until all the constraints have been moved into the critical resource portion of the network. These resources are scheduled using the Brain of OPT in a finite scheduling fashion. The remaining resources, which comprise the bulk of the resources in any manufacturing plant, are scheduled using the Serve system based on the schedules of the critical resources.

It is useful at this stage to review the OPT approach against our three criteria and nine principles. We can see that OPT clearly meets the three criteria for correct schedules.

THE OPT PRINCIPLES	THE OPT SYSTEM
1. Plant capacity should not be balanced	Balances flow not capacity
2. The utilisation of a non-bottleneck resource is not determined by its own capacity, but by some other constraint in the system	Schedules all non-bottleneck resources based on the constraints in the system. Brain of OPT feeds Serve
3. Activating a resource is not synonymous with utilising a resource	Generates schedules to maximise throughput, minimise inventory and protect the schedule from disruption – not to activate resources
4. An hour lost at a bottleneck is an hour lost of the total system	Saves setups at bottleneck operations to maximise throughput
5. An hour saved at a non-bottleneck is a mirage	Schedules additional setups at non-bottlenecks when there is sufficient idle time in order to minimise inventory
6. Bottlenecks govern both throughput and inventory	Uses inventory buffers only to protect throughput
7. The transfer batch may not and many times should not be equal to the process batch	Uses both transfer and process batches
8. The process batch should be variable and not fixed	Utilises variable process batches
9. Schedules must be established by looking at all the constraints simultaneously	Considers capacity, material and management policy constraints simultaneously

Fig. 10 Comparison of the OPT principles with the approaches used in the OPT system

- Meeting the goals of making money – the OPT schedules generate the maximum throughput with the least inventory for a given set of operating expenses.
- Realism – detailed schedules are generated that can be analysed by first line supervision to assure they are 'doable'.
- Immune to disruption – safety stock and safety capacity are employed at strategic points to prevent Murphy from disrupting the throughput of the total system.

Fig. 10 compares the OPT principles with the approaches used in the OPT system. The result is OPT's schedules are valid and followed religiously on the shop-floor.

OPT – A productivity improvement tool

If OPT is able to generate schedules that meet the three criteria of good scheduling, then it should also be a powerful simulation or 'what if' tool. It is useful to think of OPT as a system that produces the best practical schedule to meet a marketing 'wish list' under the three types of constraints (Fig. 11). The 'wish list' that we give OPT can be either firm orders, sales forecasts, or a combination of the two. We must also define three types of constraints for OPT:

- Materials – what is our current inventory position on raw materials? Are there any delivery limitations, and if so, what are the expected delivery dates or lead-times that we must live with?
- Capacity – is a variable that can be changed but at any point in time our capacity is fixed. We have so many people, fixtures, pieces of equipment, and the like, and we plan to work them a certain number of hours. It is these constraints that we must give OPT.
- Policy – these constraints can be thought of as management parameters or knobs on the OPT tool to meet their specific needs. If a company wants to be a low-cost producer, a parameter can be changed to minimise set-up cost at some expense to delivery or just the opposite if delivery is crucial.

Once the marketing 'wish list' and the three sets of constraints have been defined, OPT will generate the best practical schedule to meet the requirements without violating any of the constraints. This schedule, however, may or may not produce our 'wish list' on time. However, given the three sets of constraints, we believe this is the best possible schedule, the maximum throughput and minimum inventory, that can be obtained from our resources. If the schedule and throughput are not acceptable, then, in order to improve, we must alter one or more of the constraints.

OPT will have told us how much throughput we have got out of the system and the level of inventory. We have essentially dictated to OPT what the operating expenses are when we set the capacity constraints.

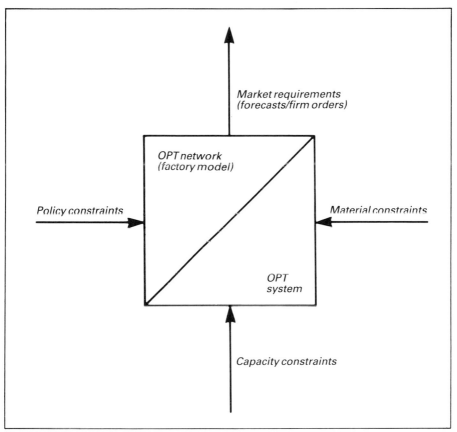

Fig. 11 OPT as a 'what if' tool

So we have defined our three measures of making money and should measure any changes in the schedule against these elements. The results of the OPT run will also show what capacity or material constraints are affecting our throughput. We can see which operations are fully loaded or where material availability impacts our schedule. If capacity is the real constraint to meeting the schedule, then we can only improve throughput by increasing the capacity at the bottleneck resources, by off-loading some of this work, or by varying our policy constraints. As one of the policy constraints is the minimum batch size that we will run, we could lower this minimum and obtain more throughput without adding additional capacity to the system.

Essentially we will be drawing down the pipeline of work-in-process inventory and converting more of it to throughput. This is an approach that may be effective in the short run but unless the capacity constraint is remedied, will not be effective in the long run. What we will have identified is where the constraints are and, as a result, we are now in a position to alter these constraints and determine the 'what if' impact of these changes. If a machine is a bottleneck, we may wish to add more overtime or make some other type of change which increases the capacity at that particular operation. We give this changed information to OPT and rerun our model. We now get a new set of outputs showing us our throughput and inventory as well as the detailed schedule at each operation. We can judge, based on this information, whether or not we have made an improvement.

OPT versus Just-in-time

An excellent comparison of OPT versus Kanban was made by Tom Murrin, President of Westinghouse's Public Systems Company, at the 1982 International APICS Conference [7]. He said: "OPT produces optimised production schedules and like Kanban it is geared to reducing inventories and identifying bottlenecks. However OPT is a computerised system whereas Kanban is a manual one. This means that bottlenecks and the impact of alternative approaches can be analysed in advance without creating problems on the factory floor. The use of men, materials and machines is optimised to maximise the utilisation of critical resources, maximise plant output and minimise work-in-process inventory and manufacturing times."

In addition, OPT can be used more universally than Just-in-time. It is applicable not only in repetitive manufacturing but also in a job shop and process industries. Finally, it has the potential to generate even greater benefits than the Just-in-time system. A key difference between OPT and Just-in-time is that Just-in-time must maintain a 'logistical chain (the Kanban) between operations where OPT has a logical one'. This difference allows OPT to be used in all types of manufacturing and eliminates most of the dams in our river system.

References

[1] Fox, R.E. 1983 OPT – An answer for America, Part III. *Inventories and Production,* Jan/Feb, 1983.
[2] Fox, R.E. 1983 OPT – An answer for America, Part II. *Inventories and Production,* Nov/Dec, 1982.
[3] Fox, R.E. 1984. OPT vs MPR – Thoughtware vs software, Part II. *Inventories and Production,* Jan/Feb. 1984.
[4] Fox, R.E. 1984. MRP, Kanban or OPT – What's best? *Inventories and Production* July/August, 1982.
[5] Goldratt, E.M. and Cox, J. 1986. *The Goal.* North River Press, Croton-on-Hudson, New York.
[6] Goldratt E.M. and Fox, R.E. 1986. *The Race.* North River Press, Croton-on-Hudson, New York.
[7] Murrin, T.J. 1982. Management for productivity and quality of Westinghouse. In, Proc. Int. APICS Conf., 1982.

SCHEDULE PROCESS MANAGEMENT

D. A. Fulghum
Navistar International Corp., USA

This paper on how schedule process management is applied at
Navistar should provide an understanding of the benefits of schedule
stability. It is recognised that truck manufacturing is somewhat
different from automobile manufacturing, but the scheduling
principles are basically the same. First, a description of how orders
are managed and how they flow through the system is given. How the
orders feed into the material requirements released to suppliers and
the things done to provide uniform plant loading for suppliers as well
as Navistar, are discussed.

Navistar has been developing scheduled process management (SPM)
practices and a means of measuring resulting schedule stability. The
company is in the relatively unusual position of being both an original
equipment manufacturer (OEM) and a supplier to other vehicle
manufacturers. For example, it builds trucks with the same engines that
are sold to other truck manufacturers. As a result of being both OEM
and supplier, the company experiences both sides of SPM. Therefore, it
is realised that every OEM and every supplier, because of the
complexity or lack of complexity of his product, has a greater or lesser
ability to meet the same levels of schedule stability. On the other hand,
the level of schedule stability attained determines the cost-saving
benefits that can be achieved from SPM.

Fig. 1 shows the flow of customer orders and their effect on
production requirements. The bottom portion of the figure indicates
'the order-board requirement', or the status of firm orders in the system
and the frequency for updating their status. Customer orders are used
to order material from suppliers. On a daily basis, orders go into a
monthly bucket (a computer term referring to a schedule time period).
Material requirements are provided to the company's suppliers either
daily, weekly or monthly, depending on their needs. There is a
minimum of seven weeks and usually two full months of firm orders.
That is, there are orders for all the trucks the company is planning to
build in the next two months. In contrast, US automobile companies

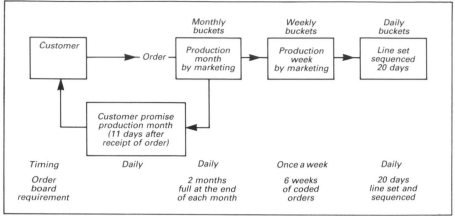

Fig.1 Order flow

run with three weeks of orders. Japanese auto makers operate with three months of orders.

Orders from the monthly buckets move into weekly buckets, once each week. Six weeks of fully coded orders are needed, which means that the manufacturing plant has all of the information needed to build the truck six weeks prior to production. Much more vehicle customising takes place in the truck business than in the automobile business, and if a customer asks for something never built before, the special features have to be engineered. As these orders come in, they are slotted in month two or beyond and referred to engineering. This department determines the special requirements and sends instructions to the manufacturing plant. At that point, the order is considered coded and is slotted into a weekly bucket.

Line-set sequencing

A 20-day sequenced line set is maintained, i.e. it is known for the next 20 days what trucks will be built each day and in what sequence. As production is completed, each day, another day is added to the line set and day one which was built is dropped. Major suppliers receive a daily printout of material requirements for the trucks planned to be built each day for 20 days, along with shipping instructions. The company's own plants that manufacture components ship within about one to three days of production. One to five days of inventory of most major components is maintained.

Again, contrasting the automobile and truck industries, Navistar runs with two months of firm orders and 20 days of daily buckets, whereas auto industry plants run on 15 days of orders. One of the reasons for this is that Navistar has about 60,000 part numbers in a plant, therefore, there is considerably more information to give to suppliers because of the complexity of the product.

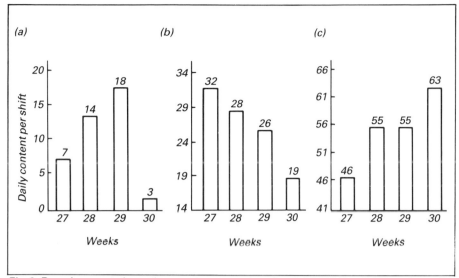

Fig.2 Requirements based on customer orders for: (a) 9670-line CAT engines, (b) 5-line 6.9L engines, and (c) 5-line 466 engines

Table 1 Example order receipt flow

Orders	Weeks								Total	
	1		2		3		4			
	S	P	S	P	S	P	S	P	S	P
Week 1	8								8	
8 standard										
2 premium		2								2
Week 2			10							10
10 premium										
Week 3					5				5	
5 standard										
5 premium						5				5
Week 4								3		3
7 standard										
3 premium							7		7	
Total	8	2	10	5	5	7	3		20	20

Assumptions: 4 weeks in a month/5-day week; total production is 40 for the month; line rate of 2 per day; two models, standard or premium.

Levelling daily requirements

The number of orders received from customers for different trucks with different options varies considerably from week to week. For example, the weekly average requirements based on customer orders for different engines are shown in Fig. 2. The company does not expect its suppliers to build and ship to these varying requirements, anymore than it could expect its lines to produce trucks to these requirements. Navistar tries to level the requirements over monthly periods to maintain efficiency for both the supplier and itself. To do this, some orders are shifted to different weeks.

A very simple example of how the production plan is levelled out is shown in Table 1. The mix of orders received on a weekly basis for standard and premium trucks is erratic over the four weeks, or 20 days, for example. If the company tried to produce to a weekly schedule based on the sequence in which the orders were received, its suppliers and itself would be operating very inefficiently from the standpoint of labour and inventory. However, viewing the production requirements for the entire four weeks, shows that there are orders for 20 standard and 20 premium trucks.

If the weekly schedule is readjusted, one of each type of truck can be produced per day. The result of this readjustment is shown in Table 2. Prior to the readjustment, suppliers were given material requirements based on orders entered as received. After readjustment, of the total 40

Table 2 Example order receipt flow after readjustment

Orders	Weeks								Total	
	1		2		3		4			
	S	P	S	P	S	P	S	P	S	P
Week 1 8 standard 2 premium	5	2	3						8	2
Week 2 10 premium		3		5		2				10
Week 3 5 standard 5 premium			2		3	3		2	5	5
Week 4 7 standard 3 premium					2		5	3	7	3
Total	5	5	5	5	5	5	5	5	20	20

% movement = 14/40 = 35%

Table 3 9670 order board realignment

Current slot	New slot week 8431	8432	8433	8434	8435	8436	Total slotted	Bal.not slotted
8431	182	59	1	–	–	–	242	–
8432	85	268	103	24	2	1	483	–
8433	3	102	244	18	2	–	369	–
8434	–	–	157	245	37	4	443	–
8435	–	–	2	115	280	69	466	–
8436	–	–	–	8	69	205	282	14
8437	–	–	–	–	–	26	26	N/A
8438	–	–	–	–	–	–	–	N/A
8439	–	–	–	–	–	7	7	N/A
Total:	270	429	507	410	390	312	2318	–
Available:	242	483	369	443	466	296	2299	14

orders 14 (35%) were moved into a different weekly bucket. All of the parts associated with the 14 orders are now expected in different time frames. The suppliers see instability in the schedule. This example reflects the trade-off that exists between level schedules and stable schedules.

Order-board realignment

An actual example of an order-board realignment is shown in Table 3 for the six-week period from the 31st week of 1984 (8431) to the 37th week (8437). The realignments took place in weekly buckets and resulted in uniform plant loading for 30 key features. The 'current slot' dates, shown at the left for each horizontal row, are the dates that the orders were placed in individual weekly buckets. The next to the last column on the right shows the total number of truck orders in each week. The weekly dates across the top of the table are the weeks where trucks will actually be produced. So, for the 31st week (8431), 242 trucks were ordered. After realignment, 182 remained in the original slot (8431), 59 moved out one week (8432), and one moved out two weeks (8433); 88 orders were pulled forward to fill the week. In the first week, 75% of the trucks remained in their original slot, and in the next week (8432), 268 of the 483 trucks ordered (or 55%) stayed in their original slot. The average percentage of trucks that did not move to another slot in this six-week example is 62%, and only 2% moved two or more weeks.

Table 4 shows the resulting average daily model and feature content after the orders were realigned to provide uniform plant loading. The far right-hand column shows the average daily content for the six-week period. The actual average daily content for each of the six weeks is shown in the respective columns. The first section reflects the vehicle

Table 4 9670 order board realignment – resulting daily model and feature content

			New slot week				
	8431	8432	8433	8434	8435	8436	
							Target average mix
Models							
A1	1.7	1.8	1.8	1.9	2.0	2.0	1.90
A2	1.2	1.1	1.2	1.1	1.2	1.3	1.19
Total	2.9	2.9	3.1	3.0	3.2	3.3	2.09
B1	67.6	68.0	67.8	67.7	67.6	68.8	67.72
B2	0.3	0.2	0.2	0.2	0.2	0	0.17
Total	67.9	68.2	68.0	67.9	67.8	68.8	67.89
C1	0	0	0	0	0	0	0
C2	6.9	6.7	6.8	6.8	6.8	6.0	6.85
C3	0.3	0.2	0.2	0.2	0.2	0	0.17
Total	7.2	6.9	6.9	7.0	7.0	6.0	7.02
Features							
1	0.6	0.5	0.6	0.6	0.6	0.8	0.61
2	16.2	16.4	16.3	16.2	16.2	16.3	16.28
3	0	0	0	0	0	0	0
4	0.3	0.4	0.3	0.4	0.4	0.3	0.34
5	0	0.2	0	0	0	0	0.03
6	0.9	0.9	0.9	0.8	0.8	0.8	0.85
7	0	0	0	0	0	0	0
8	15.3	15.3	15.4	15.4	15.4	14.3	15.34
9	7.5	7.8	8.2	8.4	8.4	8.3	8.31
10	1.2	1.1	1.1	1.0	1.0	1.0	1.05
11	19.1	19.1	19.1	18.8	19.2	19.3	19.13
12	0.3	0.5	0.2	0.2	0.2	0.3	0.27
13	3.8	3.5	3.4	3.4	3.4	3.5	3.50
14	20.4	20.4	20.2	20.1	20.1	20.3	20.27
15	48.5	48.5	49.7	50.2	49.8	43.3	48.48
16	45.6	45.8	45.8	45.7	45.8	45.8	45.77
17	11.6	11.6	11.7	11.6	11.6	11.8	11.67
18	1.2	1.1	0.8	0.8	0	0.8	0.78
Regions							
A	14.4	14.9	14.9	14.8	14.8	15.0	14.87
B	13.3	13.1	14.5	13.1	13.2	13.3	13.13
C	9.8	10.0	10.0	9.9	10.0	10.0	9.94
D	35.5	35.6	34.2	36.0	35.8	35.8	35.83
E	1.7	1.6	1.7	1.5	1.6	1.5	1.53
F	2.0	1.8	2.0	2.1	1.8	1.8	1.83
G	1.2	0.9	0.8	0.6	0.8	0.8	0.85

models' content, the second section reflects the 18 major features on the models, and the third section reflects sales regions. The production for each sales region was levelled to provide uniform distribution of product to the marketplace and to level outbound freight requirements.

   The next step in the process is to divide each week's production into daily production shift increments. Again, the requirement being that each production shift should have the same work content so that manning levels and material flow are the same from shift to shift and day to day.

Table 5 shows typical production build rates for six consecutive days. What is shown is the deviation in the daily build from the target rate. The target rate was established based on manpower and material availability during that time frame.

The final step in the process is to sequence each shift's production so that the content of each of the features is equally spaced during the shift. This provides for uniform work load and material flow throughout the shift.

Suppliers producing parts in large batches, may not care about this type of schedule realignment. Others, running complex assembly line production, on engines for example, can benefit substantially from the resulting uniform plant loading. Japanese manufacturers consider uniform plant loading to be one of the basic elements needed to achieve a high level of productivity.

Table 5 9670 daily build rate

| Restriction | Target mix | Day 1 | Build rate (deviation of target) | | | | |
			Day 2	Day 3	Day 4	Day 5	Day 6
Models							
1	2	–	–	–	–	–	–
2	69	–	–	–	–	–	–
3	1	–	–	–	–	–	–
4	2	–	–	–	–	–	–
5	0	–	–	–	–	–	–
6	4	–	–	–	–	–	–
7	2	–	–	–	–	–	–
Total	80						
Features							
1	0	–	–	–	–	–	–
2	30	–	–	–	–	–	–
3	0	–	1	1	1	–	–
4	0	–	–	–	–	–	–
5	0	–	–	–	–	–	–
6	1	–	–	–	–	–	–
7	0	–	–	–	–	–	–
8	7	–	–	–	–	–	–
9	12	–	(10)*	–	–	–	–
10	1	–	(1)	(1)	–	–	–
11	32	–	–	–	–	–	–
12	0	–	1	–	–	–	–
13	3	–	–	–	–	–	–
14	27	0.1	(0.2)	0.1	–	(0.4)	–
16	36	–	–	–	1	–	–
17	12	–	–	–	–	–	–
18	2	1	(2)	–	–	–	–

* A large customer order with this feature was not available for production in day 3

With this uniform plant loading, the company can run the same product mix down its assembly line for several weeks in a row with level parts and manpower requirements.

Supplier requirements

Various printouts are prepared for suppliers to inform them of the status of their deliveries relative to past, current and future production plans. Some suppliers receive the printouts electronically every day – a computer transmits to them during the night, and they have printouts the next morning when they arrive at their desks. Some major suppliers line set their own assembly lines from these printouts – they are running their businesses from them so a high level of credibility with stable schedules must be maintained.

Measures of how successful Navistar is in maintaining stable schedules are available. Fig. 3 is a 'build credibility' chart for a six-month period from November 1984 to April 1985. Build credibility is the percentage of trucks built in the month which the company promised both the customer and the supplier it would be built. To some extent, the credibility level is a function of supplier performance as well as the company's own. Results proved consistently above 96% and have averaged about 97.5%. The credibility performance is important to both the suppliers and customers.

When this programme started, the credibility level was in the 85% range. The target was to get above 90%. With basic SPM disciplines Navistar has been able to accomplish considerably more than it had estimated. A second measure of schedule stability used is 'line

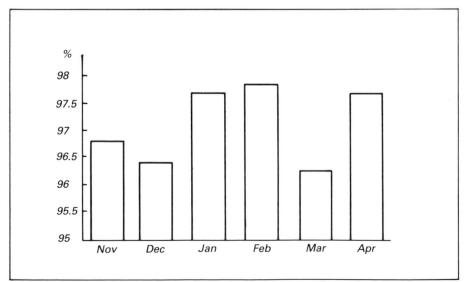

Fig.3 Build credibility

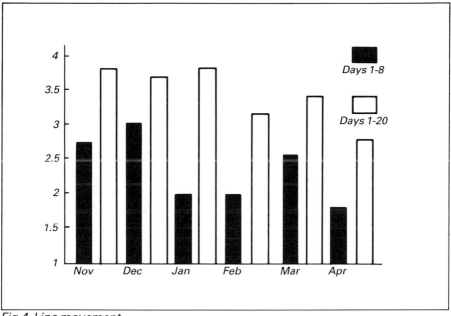

Fig.4. Line movement

movement', which is the percentage of trucks that are not built in the exact sequence that was determined 20 days prior to production. Fig. 4 shows the performance for the same six months. Line movement within eight days of production was about 2.5%, while movement in the total 20 days was 3.5–4.0%. Line movement usually results in schedule instability for the suppliers.

STOCKLESS PRODUCTION
A MANUFACTURING STRATEGY FOR KNOWLEDGE SOCIETY

M. Fuse
Sumitomo Electric Industries Ltd, Japan

Stockless production is a system of principles and methods for
realising efficient production and effective production control. This
paper presents 13 principles and eight measures developed through
simulations and actual experience.

In the 1980s, the USA, the European countries, and Japan entered the
transition era from industrial society to knowledge society. Knowledge
society is where people value knowledge, i.e. new technology and
original designs and concepts, rather than mass-produced goods and
service. This transition was made possible by two factors.

The first is that peoples' basic needs have been more than satisfied in
industrial society by its efficient mass production capability. Ordinary,
mass-produced clothes and automobiles can no longer attract the
consumers' attention. Since they demand goods and service not for
their basic, common needs, but for their individual tastes, the market
demands are divided into much smaller segments. The product variety,
therefore, is increased, while the sales for each product item are
drastically decreased.

The second factor is innovative technology currently emerging in
many fields, such as new materials, computer and communication
systems, opto-electronics, and bioscience. These new technical seeds
are giving birth to a variety of new products while making many others
obsolete. As a consequence of this, product life cycles have become
increasingly shorter.

Manufacturing firms need to adapt themselves to these changes in
order to survive in knowledge society. Stockless production is a strategy
for this. The main body of this paper was written in 1980, around the
beginning of the transition era. Although such new information
technology as distributed databases and LANs (local area networks)

now enable us to develop much more sophisticated control systems than that described in Part 2, principles and measures formulated here remain not only valid, but become increasingly crucial as knowledge society unfolds.

Drawbacks of mass production in knowledge society

Mass production developed in industrial society is no longer effective in knowledge society; if a firm still employs a mass production system, it cannot maintain market competitiveness even with its uniform quality and low-cost products, simply because these are not what knowledge society demands.

For mass production to be effective, three conditions must exist: small product variety, large order quantity, and long product lives. Based on these conditions, mass production strategy was developed with the following elements:

- *Division of labour* – To increase production efficiency while employing unskilled labour, operations are divided into many smaller and simpler pieces and assigned to many different workers. How effective this is to increase productivity is illustrated in an experiment in Part 1. Although it drastically lengthens production lead-times as also shown in the experiment, carrying finished goods inventories can offset this problem.

- *Large lot production* – Very large production lot sizes are employed in order to reduce set-ups. Production lead-times are also lengthened by this, but again finished goods inventories solve the problem.

- *Make-to-stock production* – To offset long lead-times caused by division of labour and large lot production, finished goods inventories are maintained. Make-to-stock production also enables a levelling of production; however the market demand fluctuates, a constant rate of production is maintained to maximise efficiency by the use of finished inventories as a buffer.

Mass production strategy takes full advantage of stock, i.e. work-in-process and finished goods inventories, to minimise production costs and to achieve uniform product quality. Based on the three conditions mentioned above, namely, small product variety, large order quantity, and long product lives, carrying stock was not very expensive. As knowledge society unfolds, however, the conditions are rapidly disappearing; product varieties are exploding, product lives are becoming increasingly short, and order quantities are minimised. Consequently, the mass production strategy brings forth the following drawbacks in knowledge society:

- *Increased inventory carrying costs* – When the product variety is doubled, the inventory quantity is more than doubled. Also the risk of product obsolescence increases as products' lives become shorter.

- *Increased production control difficulties* – The variety of work-in-process is increased proportionally to the product variety. More efforts are needed for materials control, scheduling, and capacity control.
- *Weak market competitiveness by long lead-times* – If a firm wants to reduce finished inventories to save costs, it faces another problem, i.e. long lead-times, which will cause lost sales and a weak market position.
- *Slow production improvements* – As new products emerge, new types of quality defectives appear. Long production lead-times, however, delay inspection and improvements in product design or process. Also machine repairs tend to be slow when the subsequent shops have an abundant work-in-process.

In knowledge society, mass production strategy is not only ineffective, but also harmful.

Stockless production for knowledge society

To adapt to knowledge society better, we must resort to a new production strategy – 'stockless production'. 'Stockless' implies that we can no longer shield ourselves by large finished inventories from the market; we need to be much more responsive to the changing market demands. 'Stockless' also implies that we must be freed from the burden of large work-in-process; we need to be much more flexible in producing a large variety of products with small quantities.

To formulate such a new strategy, it was necessary to return to the basic principles of production. They not only helped us to understand the theoretical basis of mass production, but also suggested a new strategy under the new market environment. This analysis revealed that although many approaches are proposed including Kanban and JIT, they are consciously or unconsciously based on the same principles stated in this paper. Also, one should try to respond to the same global market changes caused by knowledge society.

Therefore it is hoped that the reader will not just borrow a concrete means or a system from other firms, since the best system will be different as products and processes vary. The reader should first understand the basic principles stated here, and realise changes his firm faces from emerging knowledge society, and then build his own system to best suit his situation. If a production system is developed in this way, we would like to call it a stockless production system, whatever technical means are employed.

Part 1 – Theories of stockless production

Study group approach to develop stockless production

Stockless production is the product of two study groups: the Production Planning and Control Study Group, which began in October 1979 within Sumitomo Electric, and the IE Application Study Group, which was formed in July 1978 as one of the study groups of the Kansai Institute of Management Information Science. Each group holds monthly meetings where group members study production control through plant tours and discussions.

It is the fundamental belief of both groups that principles of production control are the same regardless of the type of product or process, although the application of the principles varies. Thus, each plant should use these principles to develop its own production control system suitable to its type of product and process.

Objectives of stockless production

By the reduction of in-process material and production lead-time, and by greater flexibility of production, stockless production achieves the following objectives:

- Revealing latent problems and weak points through reducing in-process material, thus encouraging the manager to immediately solve problems that otherwise might be put off.
- Reducing control cost by effective production control.
- Increasing flexibility to respond to demand fluctuation.
- Improving reliability of delivery-time control.

Principles of production control

In this section, principles concerning delay, in-process material, etc. are described.

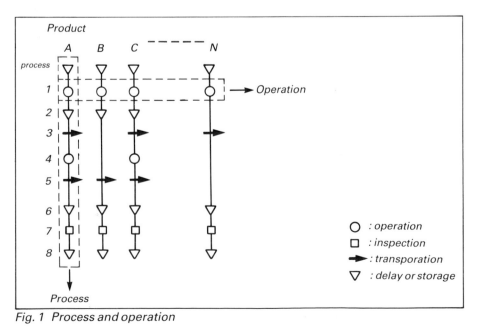

Fig. 1 Process and operation

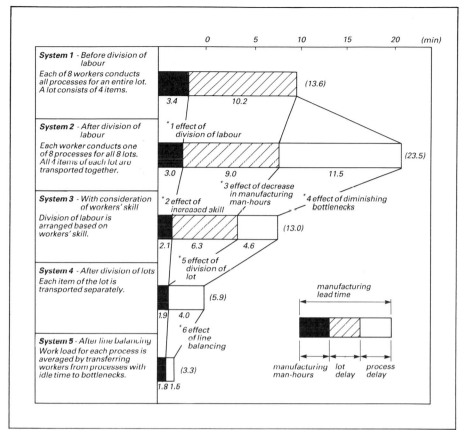

Fig. 2 Manufacturing man-hours and lead time

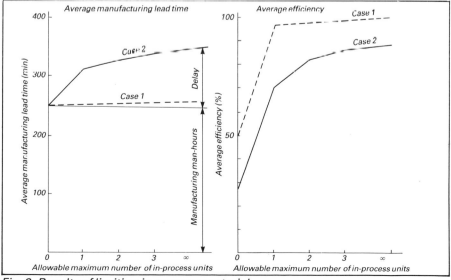

Fig. 3 Results of limiting in-process material

Two dimensions of production. There are two dimensions from which to view production (Fig. 1): one is 'process', which is the stream of goods from raw material to finished products; the other is 'operation', where materials are processed by workers and machines.

Difference in manufacturing man-hours and lead-time by type of production system. Manufacturing man-hours and lead-time to produce a specific product vary according to the type of production sytem. Fig. 2 shows the result of a simulation where five systems were compared. (This simulation was originated by Shigeo Shingo, a leading IE consultant in Japan, and is slightly modified here.)

Results of limiting in-process material. Fig. 3 shows the results of computer simulation, where in-process material is limited to specific quantities. In case 1, the line is well balanced, with each process having a cycle time of 50 minutes and fluctuation of 2 minutes. Although the line in case 2 has the same cycle time, it is not well balanced: fluctuation is as great as 40 minutes.

Balancing and timing of load and capacity. Unbalance of load and capacity is one of the major causes of delay (Fig. 4). The I/O chart (Fig. 5) is helpful to analyse the balancing and timing of load and capacity.

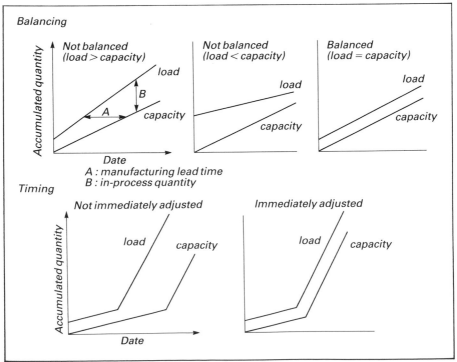

Fig. 4 Balancing and timing of load and capacity

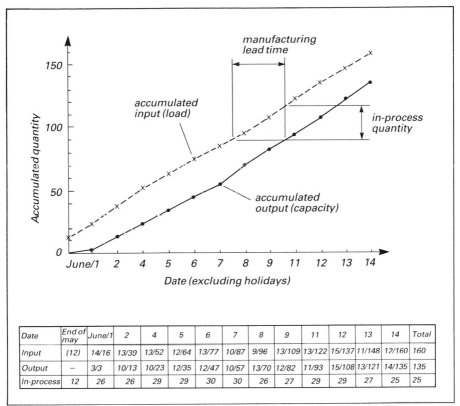

Fig. 5 I/O chart

Date	End of may	June/1	2	4	5	6	7	8	9	11	12	13	14	Total
Input	(12)	14/16	13/39	13/52	12/64	13/77	10/87	9/96	13/109	13/122	15/137	11/148	12/160	160
Output	–	3/3	10/13	10/23	12/35	12/47	10/57	13/70	12/82	11/93	15/108	13/121	14/135	135
In-process	12	26	26	29	29	30	30	26	27	29	29	27	25	25

This chart is used to analyse the work of a system which has input flow, output flow, and backlog. It shows how quantity of backlog of orders for the entire plant, or of shop orders for a specific process, varies according to changes in input and/or output. The vertical axis in Fig. 5 shows accumulated quantity of input flow and of output flow; the horizontal axis shows dates; initial quantity of backlog at the outset is shown on the veretical axis; lines showing accumulated input and output are drawn on the chart; the vertical distance between the two lines shows the quantity of backlog (manufacturing lead-time, i.e. average time period between input and output, is exhibited by the horizontal distance between these lines).

Elements of production lead-time. Production lead-time is divided into five elements, as shown in Fig. 6.

Planning and control of production. Table 1 defines 'production planning', which determines the static framework for production, and 'production control', which is the dynamic, day-to-day adjustment of production.

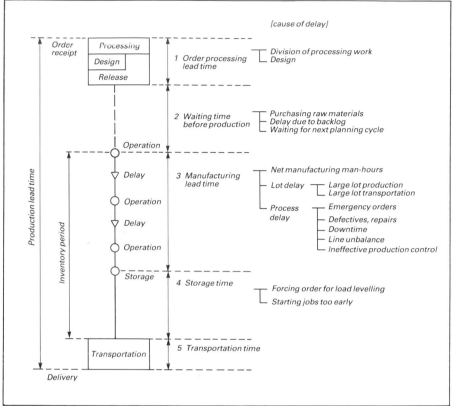

Fig. 6 Model of production lead time

Principles can be summarised as follows:

Principle 1 – There are two dimensions of production, which cross each other: 'process' and 'operation'.

Principle 2 – Delay occurs as the result of division of labour, and transportation occurs when a process is divided into sub-processes.

Principle 3 – There are two kinds of delay: 'lot delay' and 'process delay'. (Lot delay means the delay of an item while other items of the same lot are processed. Process delay occurs when a whole lot waits between processes.)

Principle 4 – The greater part of manufacturing lead-time consists of delays.

Principle 5 – Process delay is considerably reduced by balancing the line and diminishing bottlenecks.

Principle 6 – Lot delay is reduced by the division of lots.

Principle 7 – Limiting in-process material to smaller quantity is an effective means to shorten manufacturing lead-time.

Principle 8 – Balancing the line is effective in shortening of manufacturing lead-time and, at the same time, in raising efficiency.

Table 1 Planning and control

Planning	Control	
Production load levelling		

Long-range production plan
-annual (or semi-annual) plan on which
  plans of equipment, workforce, and
  finance are made
	Lead time control
	-weekly or daily adjustment of the
	  planned capacity to actual demand	

Monthly production plant
-monthly plan of quantity of each product
  group, on which necessary man-hours is
  determined		

Production scheduling		

Priority planning
-determination of delivery date and
  starting and finishing dates of key
  processes for each order (these dates are
  used as milestones for control)
	Dispatching
	-daily assignment of
	  jobs to workers and/or
	  machines, based on
	  shop schedules
	  (necessary materials,
	  parts, or tools must be
	  prepared in advance so
	  that dispatching is
	  reliable)
		Expediting
		-monitoring actual
		  performance and
		  adjusting it to the
		  planned schedule

Scheduling -allocation of orders to each shop and determination of shop schedules		

Principle 9 – Whether a line is well balanced or not, efficiency plunges to a much lower level if in-process material is totally eliminated (i.e. no delays). Thus, the optimal number of in-process units should be determined deliberately in order to balance efficiency and lead-time.
Principle 10 – Capacity should be immediately adjusted to the fluctuation of load in order not to increase lead-time (or process delay).
Principle 11 – Waiting time before production begins should be adjusted by controlling backlog.
Principle 12 – Shortening the planning cycle leads to the reduction of waiting time.
Principle 13 – Maintenance of stock of semi-finished products is an effective means to shorten manufacturing lead-time of made-to-order products. However, the stock should be kept at a minimum.

Toward stockless production

Prior to beginning the establishment of stockless production, the following policies were formulated:

- Stockless production must be established by utilising the principles of production control
- In order to reduce delay, it is not enough to deal only with delay. The overall production system must be revised.
- The target level of stockless production must be determined by the production manager.
- All personnel must participate in production control.

The measures for implementing stockless production are (Fig. 7):

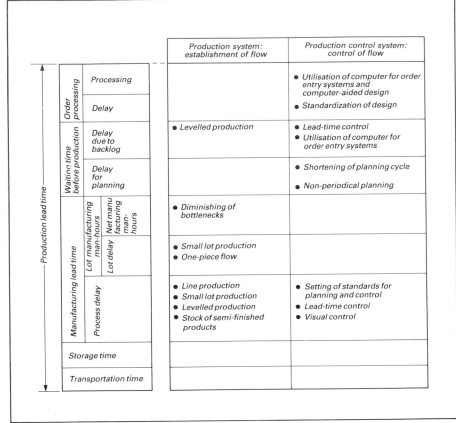

Fig. 7   Measures for shortening production lead time

- Establishment of 'flow'
  – levelled production
  – small lot production and one-piece flow
  – shortening of processes and line production
  – stock of semi-finished products
- Control of 'flow'
  – shortening of planning cycle, non-periodical planning
  – setting of standards for control (e.g. in-process quantity, flow time, etc.)
  – lead-time control
  – visual control

Measures to establish flow

Levelled production (Fig. 8). To establish flow of production and, at the same time, to adhere to deadlines, the following four measures for levelling production must be taken successively. As Principle 5

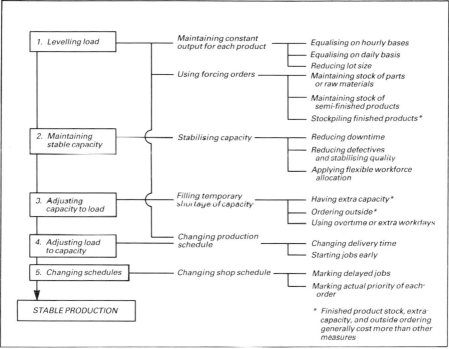

Fig. 8 Measures for stabilisation of production flow

suggests, to level load is the most important step. Levelling must be conducted for each product group, not only for the total production quantity of the entire plant:

- Maintaining constant output of each product per specific time period
- Reducing lot size.
- Using forcing orders.
- Changing production schedule.

Small lot production and one-piece flow. As Principle 6 states, lot delay is reduced by the division of lots, that is, by small lot production and one-piece flow. Fig. 9 shows the various combinations of manufacturing lot sizes and transportation lot sizes.

Shortening of processes. Principle 2 states that division of labour produces lot delay. Measures to reduce lot delay are:

- Establishing line production for each product group.
- Reducing the number of processes by integration or simplification.
- Establishing multiprocess operations for individual workers.

Stock of semi-finished products (Fig. 10). If manufacturing lead-time cannot be reduced by the measures listed previously, or if set-up cost is

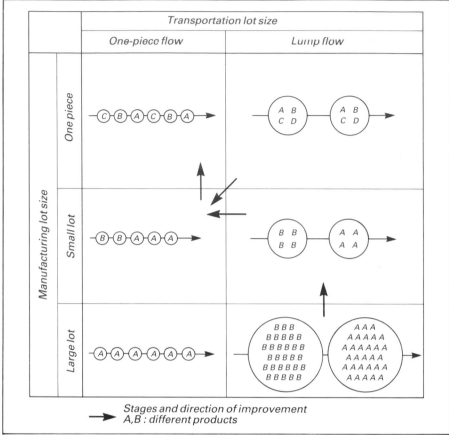

Fig.9 Manufacturing lot size transportation lot size

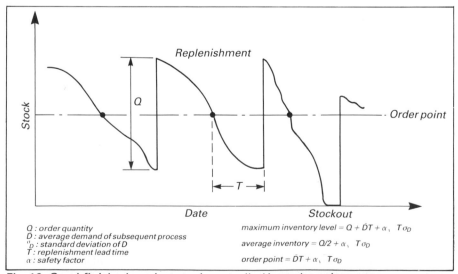

Fig. 10 Semi-finished product stock controlled by order point system

Q : order quantity
$\bar{D}$: average demand of subsequent process
σ_D : standard deviation of D
T : replenishment lead time
α : safety factor

maximum inventory level $= Q + \bar{D}T + \alpha\sqrt{T}\sigma_D$

average inventory $= Q/2 + \alpha\sqrt{T}\sigma_D$

order point $= \bar{D}T + \alpha\sqrt{T}\sigma_D$

too high to conduct small production, stock of semi-finished products, as mentioned in Principle 13, can be maintained. To control semi-finished product stock, a 'standard in-process quantity system' is used. In case of an order-point system, one type of standard in-process quantity system, the following measures are taken to reduce inventory level:

- Using smaller order quantity (Q).
- Shortening replenishment lead-time (T) by reducing order quantity.
- Minimising fluctuation in demand (σ_D) by levelling production in the subsequent process.

Measures to control flow

Flexible schedule planning. According to the method and timing of determining what, when, and how much to make, there are various types of schedule planning, as shown in Fig. 11. Table 2 divides them into four major types. Two effective means of obtaining flexible planning are: shortening of planning cycle, and use of non-periodical planning.

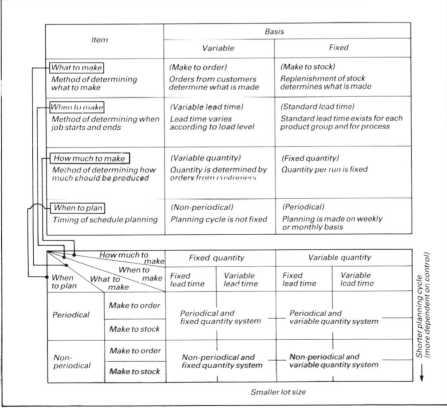

Fig. 11 Items of schedule planning

Table 2 Four types of planning

System	Periodical and fixed	Periodical and variable quantity	Non-periodical and fixed quantity	Non-periodical and variable quantity
Explanation	An order of fixed quantity is placed at a fixed interval	Production quantity is planned monthly or weekly based on backlog and/or demand forecast	A production order is placed when inventory level goes below an order point	When and how much to make is planned for each order separately
Applicability to make-to-stock:	○	○	○	
make-to-order:		○		○
Characteristics	• Production quantity must be stable	• Lead time requested must be longer than planning period	• Manufacturing lead time must be short to keep inventory levels low	• Variety in lot sizes must not be large
	• Order cycle and delivery cycle are fixed	• Sufficient amount of orders to fill planning period must exist	• Production level must be stable	• Production level must be stable
		• There can be little addition or change of plans once made		
Daily scheduling need for:	No process	Key processes only	Last process only	First process and key processes

Setting of standard – planning and control complementing each other (Fig. 12). To control the flow of production, some standards must be used. Flow speed can be controlled by standard flow time, or standard in-process quantity. Flow quantity can be controlled by standard manufacturing quantity per day.

Lead-time control. Lead-time control consists of measures to adjust capacity to load and load to capacity, as shown in Fig. 8. In order to adhere to 'standard flow time' or 'standard in-process quantity', lead-time control is necessary. There are two types of lead-time control:

- Control of manufacturing lead-time for each process (Fig. 13), e.g. control of backlog quantity of shop orders.
- Control of waiting time before production begins, e.g. control of backlog quantity of orders for the entire plant.

Visual control. To achieve participation of all personnel in production control, visual control is effective. It makes each worker aware of the

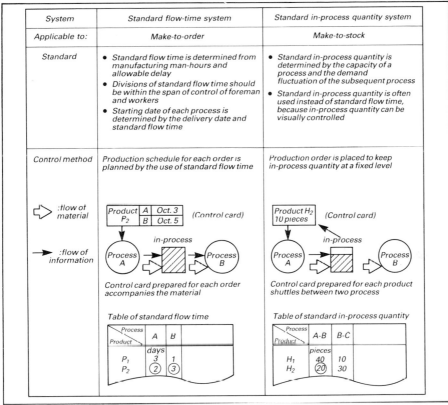

System	Standard flow-time system	Standard in-process quantity system
Applicable to:	Make-to-order	Make-to-stock
Standard	• Standard flow time is determined from manufacturing man-hours and allowable delay • Divisions of standard flow time should be within the span of control of foreman and workers • Starting date of each process is determined by the delivery date and standard flow time	• Standard in-process quantity is determined by the capacity of a process and the demand fluctuation of the subsequent process • Standard in-process quantity is often used instead of standard flow time, because in-process quantity can be visually controlled
Control method	Production schedule for each order is planned by the use of standard flow time	Production order is placed to keep in-process quantity at a fixed level

Fig. 12 Two types of control systems

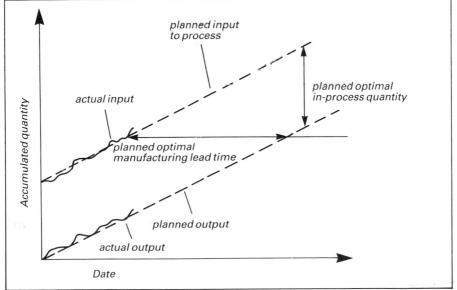

Fig. 13 Control of manufacturing lead time

Types and basic patterns	Key points	Objectives
Expediting board 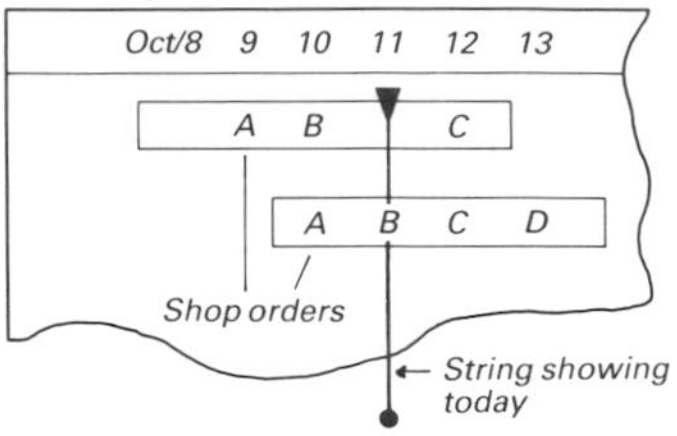 Oct/8 9 10 11 12 13 A B C A B C D Shop orders ← String showing today	• Key process to be controlled should be selected • Starting dates for each key process are shown • A string is used to show the current date	• To expedite shop orders • To visualise the gap between planned schedule and actual operation • To determine delivery date
Loading board 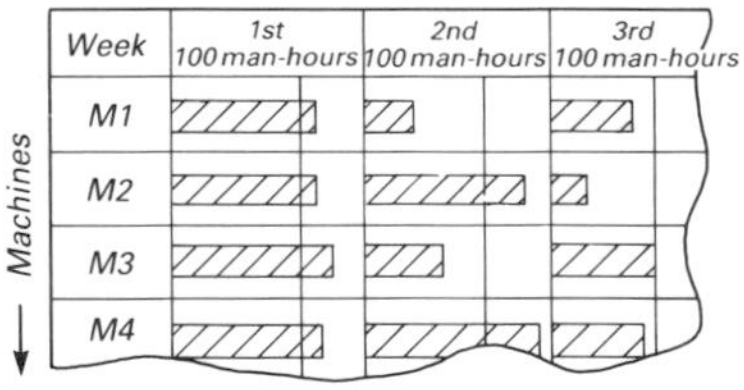 Machines Week | 1st 100 man-hours | 2nd 100 man-hours | 3rd 100 man-hours M1 M2 M3 M4	• Work loads for each week allocated to machines are graphically shown	• To make workforce plans • To determine delivery date
Dispatching board 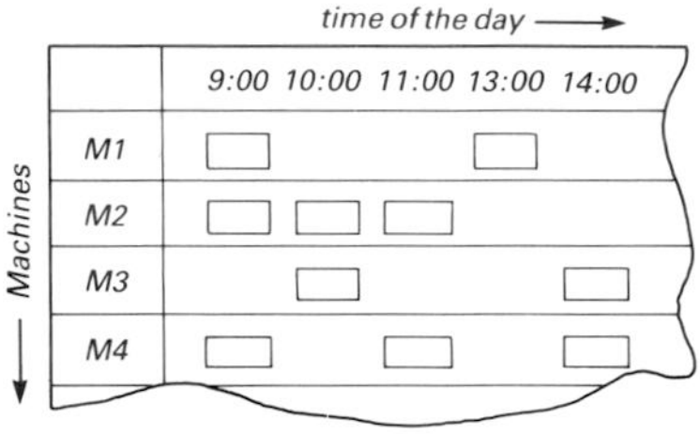 time of the day → 9:00 10:00 11:00 13:00 14:00 Machines M1 M2 M3 M4	• Shop order cards (or job tickets) allocated to each machine are put on the board • Hourly-basis dispatching is desirable	• To show present dispatching of orders for each machine
Delivery board 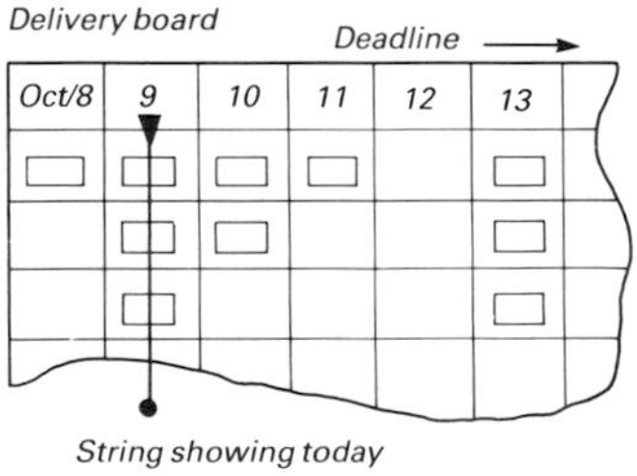 Deadline → Oct/8 9 10 11 12 13 String showing today	• Shop order cards are put on the board according to their deadlines • A string is used to show the current date • Delivery board can be used for check of supply of materials or parts for a process	• To take preventive measures for delay

Fig. 14 Four types of production control boards

current condition of production flow and enables him to adjust his own work accordingly. The following can be used:

- Visual control of materials.
- Visual production control board (Fig. 14).
- Trouble display devices.

Flexible production system

In addition to the establishment of a production control system, the following improvements are essential to assure stable flow of production:

- Shortening of set-up time (quick set-up is essential for lines with a great variety of products).
- Equipping workers with various skills, which makes establishment of a flexible workforce allocation system possible.
- Quality assurance within each process by the use of fool-proof devices and total inspection.
- Establishment of full-time production improvement teams to accelerate improvement.
- Use of multipurpose machines and tools for production lines with a great variety of products.

Evaluation of flow

Flow value. The concept of 'flow value' was developed to evaluate the rate of production flow. It is universally applicable to all products and processes:

$$F = \frac{L}{M}$$

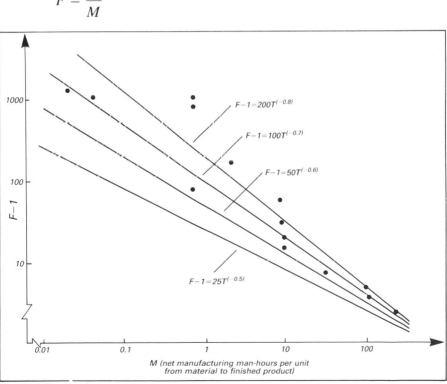

Fig. 15 *Flow value*

Table 3 Levels of production flow

Levels	Coefficients		F value
	A	α	
Ordinary level	200	0.8	$200M^{-0.8} + 1$
Small lot production level	100	0.7	$100M^{-0.7} + 1$
Stockless production level	50	0.6	$50M^{-0.6} + 1$
One-piece flow level	25	0.5	$25M^{-0.5} + 1$

where F is the flow value (or F value), L is the average manufacturing lead-time from material to finished product (number of days including holidays $\times$ 24 hours), and M is the net manufacturing man-hours per unit from material to finished product.

Levels of production flow. From 15 observations of F values, a hypothesis was obtained: the relationship between F value and M can be formulated as:

$$F = A \cdot M^{(-\alpha)} + 1$$

Fig. 15 shows this equation and the 15 observations. The four lines in Fig. 15 exhibit four levels of production flow chosen for evaluation and classification, the coefficients α and A of which are shown in Table 3.

The manager can evaluate the flow level of his production line by calculating its F value. He can also set the target value of manufacturing lead-time by his M value and the goal level of his production flow.

Part 2 – An application of stockless production

Characteristics of alloyed aluminium wire plant

Sumitomo Electric's alloyed aluminium wire plant, a subdivision (with only 30 workers) of the Osaka Aluminium Wire & Cable Plant, produces a great variety of aluminium wire products with various kinds of alloys and wire sizes, the total production quantity being 200–300 tons per month. The plant has the following characteristics:

- Customer orders
  - the number of orders is 250–300 per month
  - most orders are repeat orders, though there is a great variety of products
  - lead-time is as short as 2–3 weeks
  - small quantity orders (less than 500kg) account for a large percentage
  - rush orders and changes in requested delivery date, quantity, and wire size after receipt of the order are frequent
- Production system
  - the processes consist of two to three stages, excluding inspection and packing

- some products are partly processed by a subcontractor
- all production is done on the make-to-order basis, with no finished product inventory being kept
- the plant uses 18 kinds of raw aluminium wire; replenishment is made twice a month based on predicted requirements

Situation before improvement

A veteran expediter did all the work to adhere to deadlines. Owing to his skill, delivery delay was not very serious, but nevertheless there often appeared inefficiency from idle time because of material stockout or little work load and from overloading. On the other hand, since the demand fluctuation was too large to forecast correctly and since lead-time was very short – 80% of customer orders required delivery time within two weeks – a flexible production control system was desired.

To establish such a system, a production control improvement team was formed, just at the same time as a minicomputer terminal was introduced into the plant office.

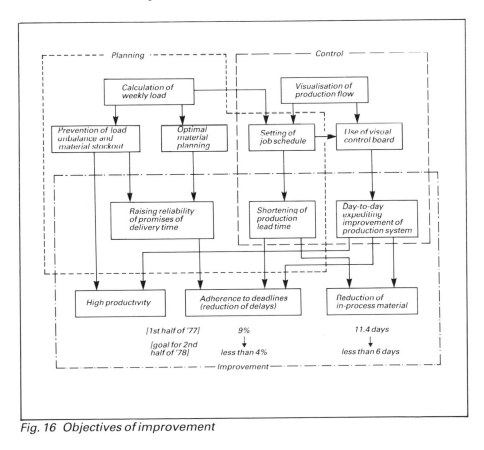

Fig. 16 Objectives of improvement

Objectives and methods

From discussions and analysis of situations by the team members, objectives were established, as shown in Fig. 16. The following were defined as basic approaches:

- Setting job schedules, which can be used as the basis of control.
- Calculating weekly work load and determining when, what, and how much to make accordingly.
- Using visual control boards to make everyone aware of present state of production flow.

The goals of improvement were set for the second half of 1978 as delivery delay of less than 4%, and in-process material of less than six days.

Slogans were made to give clear understanding of the improvement to all the personnel concerned:

- 'Participation of all in production control' (everyone should be aware of the present flow of production).
- 'Think as you proceed' (rather than spending a lot of time discussing what to do but not actually doing it, try out your new ideas, evaluate their results, and correct them if necessary).

Production planning and control system after improvement

Fig. 17 shows the production planning and control system after improvement. Steps 1 to 12 in the figure are as follows:

Step 1 – Orders are transmitted from sales divisions to the plant.
Step 2 – Delivery date for each order is determined by the scheduler based on customer's requested delivery date and, at the same time, current machine load. If an order quantity is too large, it is divided into smaller lots, since too large a lot affects production.
Step 3 – An order with its delivery date already determined is input into a computer for computation of schedule, machine load, and material requirements. (In the case of alloyed aluminium wire, the type of processing required, the machine load, and the material requirements for a particular order can be determined by a simple calculation.)
Step 4 – A production schedule, that is, the starting date for each process, is calculated backward from the delivery date using standard flow time, which has been predetermined for each process according to product type and lot size.
Step 5 – The machine load report is updated daily, using the schedules and required manufacturing man-hours for newly received orders. The report shows each machine's load for the next four weeks and is used for workforce planning and delivery

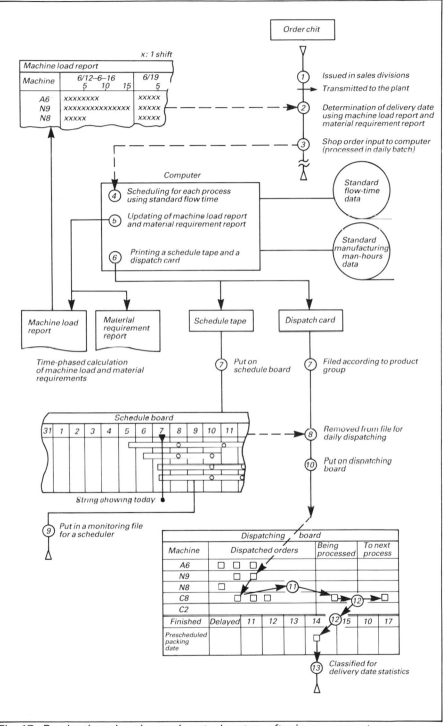

Fig. 17   Production planning and control system after improvement

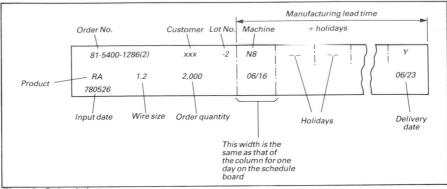

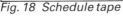
Fig. 18 Schedule tape

date determination. The material requirement report (which has an eight-week range) for replenishment of stock of raw aluminium wire is also updated and printed daily.

Step 6 – A schedule tape and a dispatch card are printed for each shop order to be used for visual control of production flow.

Step 7 – The schedule tape (see Fig. 18) is put on a schedule board. The length of the tape shows the planned number of days required for processing, including holidays.

Step 8 – A red string hangs on the column of the current date. Processing should be started for orders whose tapes are crossed by the string. Dispatch cards (see Fig. 19) of these orders are put on a dispatching board every morning.

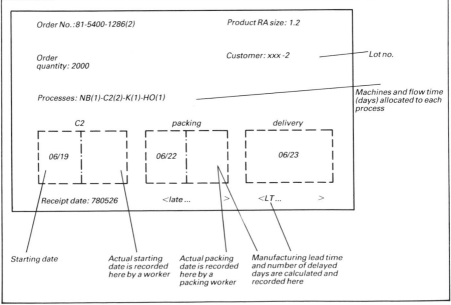

Fig. 19 Dispatch card

Step 9 – When the order starts the final process, its schedule tape is removed and filed for use in monitoring by the scheduler.

Step 10 – Workers check the dispatch board to determine which orders to process.

Step 11 – The dispatch card is moved to 'being processed' when processing is started, and then to 'to next process' or the 'finished' block when it is completely finished. As each process is finished, the date is recorded on the dispatch card.

Step 12 – Dispatch cards are returned to the scheduler after packing for use in recording delay statistics.

Characteristics of the new system

The characteristics of the new system are summarised below:

- *Flexible, non-periodical planning* – Monthly planning and frequent rescheduling were abolished. In the new system the schedule for each order is determined as soon as the order is received. Processing priority, the basis for planning and control, is determined based on these schedules.

- *Standard flow time system* – Starting dates for each process are calculated backward from delivery dates using standard flow time rather than being calculated from daily load fluctuation of each machine. (Work load has already been taken into account when setting the delivery date.)

- *Capacity planning* – A machine load report and a material requirement report are used for planning for production level, work force and material supply. The difference between planned capacity and actual demand is adjusted by the following measures:
  - using the machine load report and the schedule board, the delivery date is determined so as to level the load
  - subcontracting and division of large-quantity orders are used for load levelling
  - workers themselves can determine how many overtime hours are necessary by consulting the schedule board
  - use of standard flow time leaves room for adjustment because it includes allowance time

- *Visual control* – Control boards, designed and made by workers themselves, give workers visual information on load, dispatched orders, and present condition of production flow.

- *Use of computer* – A computer is fully used to save man-hours for calculation of each order's start dates and each machine's work load and for printing of cards and tapes. Such a simplified use of the computer makes workers familiar with the role the computer can play in a plant.

Results

The goals have been achieved, as shown in Fig. 20. Though the production quantity rapidly expanded in this period, there appeared significant improvement in the late delivery rate and in-process material quantity. Especially for product A, there was no late delivery in December 1978 or January 1979. It is assumed that the new system, which established rapid flow of production, also contributed to productivity. As another by-product, harmony among different groups was reinforced.

The new system works remarkably well and has become indispensable for the operation of the plant. Inquiries and follow-up on delivery dates, which were dealt with by a production planner before, are now handled well by the scheduler (formerly called an expediter). Workers willingly assume the responsibility of adhering to delivery dates.

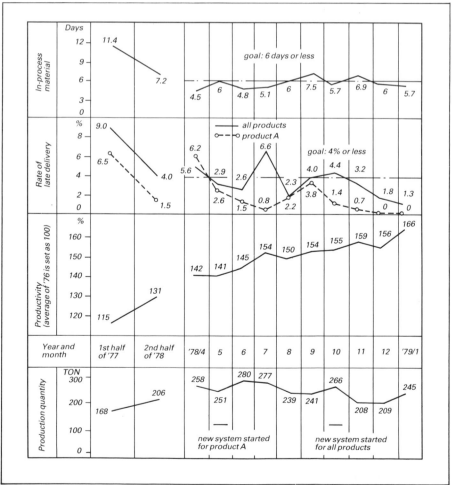

Fig. 20 Results

Table 4 Results of inventory reduction

Divisions	A	B	C	D	E	F	G	H	I	J	K	Total
Reduction rate compared to 2nd half of '77	−5%	+6	−5	−33	−30	−11	−26	−6	−33	−22	−8	−13

Concluding remarks

Inventory reduction

Sumitomo Electric engaged in company-wide inventory reduction activities in 1978. The results are shown in Table 4. As can be seen, the total target of a 10% reduction was achieved. The results of divisions D, E, G, I and J are noticeable. Each of these divisions (except G) tried to establish stockless production suitable for its own characteristics and succeeded, though not yet completely, in doing so. Their production systems and production control systems have been sufficiently improved to keep inventory levels from going up again.

Knowledge obtained through implementation of stockless production

The methods used by the above divisions to establish stockless production were closely examined, and the following knowledge was extracted:

- In many divisions, monthly planning is not able to respond rapidly enough to the large fluctuation in demand. In such cases, non-periodical planning is effective.
- Improvement only in the production control system cannot produce sufficient results. The production system itself must also be improved by using line production, levelled production, and small lot production.
- Visual control is quite effective to realise participation of all personnel in production control.
- The use of standards, such as standard flow time or standard in-process material, is also effective. It serves to make planning and control complement each other, and it enables each worker to control his own process.

Stockless production is not yet fully developed. However, it is destroying previously accepted ideas for high productivity such as large lot production, periodical planning of production, fixed workforce allocation, and pursuit of efficiency in each process independently. It is hoped that, through trials in many plants, the development of stockless production will be completed as the principal means for overall improvement of the production system.

4

ORGANISATION FOR CHANGE

This section focuses on the various considerations that companies must look at when organising for and implementing JIT. The first area is quality management. It is generally agreed that high levels of quality are prerequisite for JIT. The papers in this section look at the relationship between quality and JIT, and how the Japanese manage quality in a JIT environment. Second, JIT implementation involves two important considerations: a process of continuous improvement, and the involvement of the workforce. Finally this section looks at how JIT can be implemented. A framework is provided for the implementation of JIT and useful guidance is given on the important question of where and how to start JIT and in what order it should be implemented.

SQC AND JIT PARTNERSHIP IN QUALITY

S. Priestman

Hewlett-Packard's Computer Systems Division has discovered the synergy that exists between statistical quality control (SQC) and Just-in-time manufacturing. This paper is based on interviews with Hewlett-Packard employees.

Many companies today are approaching quality as a way of changing their employees' perception of the manufacturing process. If a company teaches quality in this broad way, it can combine complementary quality-related programmes to achieve much greater improvement than individual programmes could produce by themselves.

Hewlett-Packard's Computer Systems Division has taken this approach. In 1982, it introduced a statistical quality control programme; in March 1984, it launched Just-in-time manufacturing. Having already established statistical control of its manufacturing processes, the company found that the transition to JIT was less a problem-solving exercise than a solution-oriented project.

Rick Walleigh, production manager for the Division, saw the results of the synergy between SQC and JIT. First, the cycle time in the printed circuit assembly area dropped from 17 days in July 1982, to six days by March 1984, through the use of SQC. Then, upon implementation of JIT, cycle time dropped to 1.6 days after two months and remained at that level. Says Walleigh, "Going into JIT was simplified by the data available through SQC. We were able to target problems and establish quality before moving into a new system. Getting the quality right before introducing JIT is the key to a successful JIT programme."

The practice of JIT reflects its title: produce and deliver products just-in-time to be assembled into finished goods, fabricated parts just-in-time to go into subassemblies, and materials just-in-time to be transferred into fabricated parts. According to Schonberger[1], inventory is the root of much manufacturing evil. Walleigh agrees. "We traditionally thought of inventory as a good thing because it provides a buffer against a lot of problems and isolates one department from things occurring in other areas, so that departments can concentrate on their own areas. But it also covers up problems, which then go unsolved. Inventory hides late delivery, bad parts from vendors, procedures that constantly break down because of poor maintenance, and processes that operate inconsistently."

JIT exposes these problems because of its hand-to-mouth mode of operation. Traditionally, most manufacturing systems operate in a batch-oriented mode, which recognises a set-up time and cost. The formula used to calculate these costs is known as the economic order quantity. It is figured by balancing the expense of setting up a batch against the cost of covering materials in order to determine the economic lot size. Batches are large to amortise costs. JIT, on the other hand, fixes the lot size, and then determines the set-up cost. Rather than accepting the set-up cost as a fixed number, JIT works on lowering it. This requires a shift from a batch mode to a mode more like an assembly line.

Ingrained in the batch-oriented mode is the 'push' method of manufacturing. Each operation has a schedule and produces a batch of material according to this schedule whether the next operation needs it or not. JIT introduces the concept of demand 'pull', where the schedule is created by the final operation. When one station has completed all available work, it signals the next station to give it more; or, if an area of production is full, the operation that feeds it doesn't produce until receiving an indication – through a card or an empty space – that there is a need to produce. In practice, an operation takes lead-time into account in preparing work for the next station.

This system immediately uncovers equipment breakdown. In a push system, by contrast, material waits in piles around the malfunctioning equipment while operators continue work, processing other material according to the original batch schedule, at individual stations. They do work put aside for late material delivery; they rework non-conformities; they do work they have hidden in preparation for these down times. In a push system, the problems that caused the machine breakdown, the late material delivery, or the non-conformities would not be recognised as contributing to the quality (or the lack thereof) of the manufacturing process.

Let us look at what happens in a JIT system when equipment breaks down. When one station does not receive work from the preceding station because of breakdowns, not only is the machine not working, but the operators are also not working. Addressing the source of equipment malfunction is now critical.

If an operator does run out of work, he or she may help solve the problem causing the stoppage, or may be assigned to work in another area.

Because JIT removes the buffer of inventory accrued by traditional stocking methods, it is vital to know what the true needs of production are to order accurately. Developing an understanding of those exact needs is the function of SQC. SQC and JIT offer complementary means of reaching quality objectives; used together, both programmes are more successful.

In general terms, SQC can be broken down into three major elements[2]:

- Process analysis to understand the system.
- Inductive reasoning to measure the system.
- Leadership to change the system.

SQC allows a company to understand its operations as a set of processes. The natural behaviour of each process is defined and monitored by means of control charts. Process variations are recognised and evaluated. Non-conformities and errors can often be reduced by making changes in a process. When declines in productivity and quality appear, SQC exposes the patterns and locations of problems in processes and provides the tools to control those areas. By controlling errors, a company may also simplify its processes.

An example of SQC simplifying procedures and smoothing the introduction of JIT comes from the materials management function. For Chuck Cheshire, materials manager at the Computer Systems Division, the JIT programme meant bringing materials to the back door, just-in-time. "JIT only works if there are no back orders," explains Cheshire. "When we were initially considering JIT, we assumed that if we had more material, back orders would decrease. However, by using SQC, we discovered that this assumption was false. We had several months of inventory, yet our back orders were running at over a hundred. That told us we had the wrong inventory. We focused on the problem, did a Pareto analysis, and asked ourselves, 'What are the real processes? What are the technologies, the commodities, the vendors?'"

"We worked on these problems one at a time, and discovered a number of causes. Most were easy to fix, once the data was available. Our inventory dropped from two-and-a-half months to one month, including raw material, work-in-progress, and other finished goods. Our back orders are now down to two or three."

This means improved relations with vendors as well. Continues Cheshire, "In developing a single source, you must help your supplier be successful by providing information on the product before establishing JIT. That's where SQC comes into play. We were able to provide our supplier with specific feedback on what our incoming quality inspection found compared to the quality they defined when they sent us the parts. We also collected data on the production floor and provided the supplier with this information. Without SQC, we could only say, 'Here's the part, it's broken.' Now we can focus our efforts and the supplier's efforts to pinpoint material problems."

According to Walleigh, SQC and JIT work well together because they share four common action principles:

- Simplicity.
- Waste elimination.
- Exposing problems.
- A climate for continuous improvement[3].

Simplicity

Both JIT and SQC simplify the manufacturing process. In SQC, obtaining accurate evaluations of data to determine the source of non-conformities requires that operations be performed consistently. A major enemy of consistency is complexity, such as non-conforming raw materials, late deliveries of components, erratically functioning equipment, and poorly documented procedures. By using SQC analysis to expose and eliminate these complexities, a company can make its process consistent, reliable, and simple.

JIT simplifies the manufacturing process through material movement. Complex production scheduling of multiple levels of product subassemblies is eliminated, because when material is needed it is pulled from the preceding production operation. There is no need to track progress against schedule for the production of subassemblies because nothing is built until it is needed. An obvious signal or card displays the need. Only final production complexities need to be tracked. Employees do not have to guess at what they should work on next: the next action to be taken becomes obvious, and priorities are set automatically.

Waste elimination

As non-conformities are minimised through SQC, less scrap is produced, and labour is saved through a reduction in rework. Labour is also no longer wasted on improvised procedures with which workers previously adapted to the problems of a complex environment.

Says Perry Gluckman, SQC consultant, "Workers develop informal ways to get their work done no matter what the formal system is. But this is not the worst thing that can happen. It is far worse when the workers follow policy and build something to specification even though – because of a lack of understanding about the processes – the specification is wrong. Unfortunately, the worker is often blamed for the lack of quality in these situations."

With SQC and JIT, the workers' creativity is channelled to more productive projects. Bob Tellez, the section manager for Printed Circuit Assembly, is working with line operators to improve job satisfaction. "We've given a group of people full responsibility and ownership for major areas," he explains. "One group on day shift has responsibility for hand-loaded components on boards; they also oversee the wave-solder process. Ten people work within these two processes, but within that system, they rotate from one station to the next. When one station slows down, they can go help another. This is what the operators chose to do when we installed the JIT system, as the need for variety became evident. SQC offered the operators visible recognition for their abilities. Operators are now given opportunities to work together to make changes that will yield improvements."

JIT approaches waste from another perspective. Waste is primarily eliminated in inventory. Because products are not assembled until needed, inventory is lower throughout the manufacturing process. The obvious and direct benefit is increased asset productivity. However, the indirect benefits prove even more valuable. With less inventory separating different production operations, the cycle time is shortened and manufacturing is consequently more responsive. Another benefit is the elimination of wasted space previously used to store the unnecessary work-in-process inventory.

Because JIT requires that material be produced and moved in small amounts rather than in large batches, machine set-up times are reduced, which also eliminates waste of labour and capital equipment.

Exposing problems

SQC distinguishes between variations in a process as a result of outside causes, and variations that are the result of inconsistencies within the process itself. Once this distinction is understood, appropriate resources can be committed to eliminate the source of the problem.

JIT exposes problems in a less direct but often more compelling way. It strips away the mask of inventory that hides the problems of late deliveries, non-conforming raw material, processes that produce scrap, processes with unbalanced capacities, and poor production planning. Without the buffer of back-up inventory, JIT forces managers to recognise and address these problems.

A climate for continuous improvement

Once SQC and JIT become a part of the manufacturing routine, continuous improvement becomes everyone's objective.

SQC and JIT have produced impressive results at the Computer Systems Division. The production process has been simplified, dropping from 26 steps to 14. Employee responsibilities are clearer, and communication has improved. According to Tellez, "SQC offers everyone a sense of ownership and allows worker participation at all levels of process improvement. Implementing JIT is a simpler transition than it would be without that sense of involvement in place."

Waste and problem visibility have been targeted in a number of areas. Since the July 1982 implementation of SQC, direct labour dropped from 62 hours per product to 52 by the first quarter of 1984. Seven months after JIT was introduced in March 1984, labour hours dropped to 39. Non-conformities in pretest IC insertion products, already monitored by SQC, plummeted from 1950ppm in January 1984 to 600ppm three months later. After JIT, non-conformities fell to 210ppm by October. Hand-loading non-conformities dropped from 120ppm in January to 90ppm in March to 48ppm in October.

Solder non-conformities in the wave-solder process have demonstrated the most dramatic decline. Upon installation of SQC, non-conformities were recorded at 5200ppm. Five months later, documentation shows 100ppm in the same wave-solder process. Though fluctuations occurred during 1983 and early 1984, non-conformities were lowered to under 100 within two months of JIT installation, and have remained under control ever since.

In the wave-solder section of Printed Circuit Assembly, functional non-conformities fell from 14ppm in January to 11ppm in March to 5ppm in October 1984. PC Assembly final assembly non-conformities started at 145ppm in January, fell to 97ppm by March, and were down to 10ppm in October.

These reductions were immediately felt in the working environment. Says Tellez, "When we eliminated a lot of the wasteful conditions that existed in the batch-oriented material flow system, people were relieved. They now know that the material they will be working on is just getting finished, and as soon as it arrives, they'll be ready for it. In the old system, people didn't feel comfortable unless they had a rack of work behind them. People would hide work so they'd have it to keep them busy for an entire week. We reduced floorspace by 32% when we eliminated this backlog."

Concluding remarks

Implementing both SQC and JIT requires the commitment of management and the direct involvement of operators. In making the decision to establish quality programmes, warns Cheshire, "Management commitment is not just a matter of saying, 'you'll implement SQC, thank you.' Management commitment means allowing workers to try and, if they fail, to learn from these failures. Ten years ago, standard engineers might be on the floor, watching how much time each assembly would take. Now, through SQC, production workers know if they're in or out of statistical control; and if they're out of statistical control they know they can stop and ask for assistance from engineering.

"SQC and JIT allows managers to listen to the operators, tap into their insights, and make the system more effective while making the operators' jobs more interesting. Quality commitment supports individual initiative. If people have pride in their jobs, and are trained to control their jobs, they do well, and the company does well."

References

[1] Schonberger, R. 1982. *Japanese Manufacturing Techniques: Nine Hidden Lessons in Simplicity*. The Free Press Division of MacMillan, New York.
[2] Grant, E.L. and Leavenworth, R.S. *Statistical Quality Control*.
[3] Walleigh, R. *Synergy in TQC and JIT*. Internal report, Hewlett-Packard.

AN ANALYSIS OF JAPANESE QUALITY CONTROL SYSTEMS

S.M. Lee
University of Nebraska Lincoln
and
M. Ebrahimpour
University of Rhode Island, USA

This paper presents an overview of the Japanese quality control system and its key features and characteristics, while comparing them with the US standards. This discussion leads to the presentation of important implications for American manufacturing firms for improving their product quality.

"Hundreds of Boeing 767s have been grounded." "Thousands of automobiles have been recalled." "Thousands of TV sets have been modified." These news headlines are examples from a long list of US products with quality problems. They help confirm a general opinion among consumers that the quality of American made products does not measure up to that of Japanese made products. Therefore, it is not surprising that more Americans are buying more Japanese made products.

The change in purchasing behaviour of American consumers has cost the US economy a great deal. For instance, a trade deficit of $13 billion in favour of Japan would cause a loss of at least 600,000 jobs in the USA. In addition, for every Japanese car imported to the USA, approximately $1750 in tax revenue is lost. The US trade deficit hit a new record in 1986 – $175 billion – and the bilateral deficit with Japan was approximately $50 billion.

American manufacturers have yet to respond to Japanese competition with improved product quality and productivity. As a result, in many industries the USA no longer produces the best or the most reliable goods. US products seldom have competitive prices, especially in industries with stiff international competition such as steel, automobiles, TVs, cameras, VCRs and microprocessor chips.

Japan has emerged as the USA's number one competitor. Within three decades the Japanese changed their products' image from being shoddy and cheap to being very good quality. Today, many Japanese manufacturing firms are well known for producing the best quality products with a competitive price available to consumers. Because of this, both American business executives and academics have been trying to find the cause behind Japan's success in producing high quality products at competitive prices. Among many factors scrutinised by scholars, Japanese management style has been the centre of most discussions.

In comparing Japanese and American styles of management, various aspects such as lifetime employment, quality of working life, culture, and more recently, manufacturing techniques, have been discussed. Among the different types of manufacturing techniques, total quality control (TQC) is most familiar to Americans. TQC has accounted for great contributions to higher productivity and better quality of Japanese products. However, TQC appears to be only one part of the Japanese quality control system.

Comparative analysis of quality control systems

The modern concept of quality control was introduced in Japan after World War II, about 25 years after it appeared in the West. However, the Japanese utilised the idea better than its originators did. To achieve higher quality products, the Japanese began learning modern quality control concepts from either consultants from the West or through observation and analysis of Western quality control technology. According to W.E. Deming[1], four forces converged to cause the

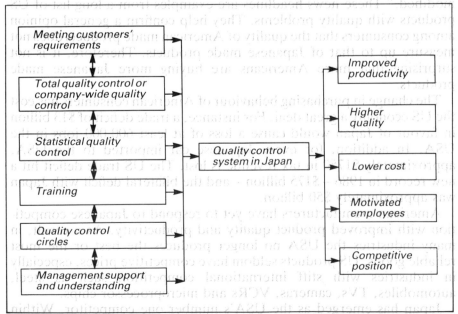

Fig. 1 *Characteristics of Japanese quality control and their potential benefits*

emergence of concern for quality in Japan. These forces were Japanese statisticians, the Union of Japanese Scientists and Engineers (JUSE), the teaching of techniques, and conferences with top management.

Although the Japanese learned the concepts relating to quality and quality control from the West, they modified and blended these lessons to fit their own organisational environment. These changes and modifications provided special features to the Japanese quality control system, such as meeting customers' quality requirements, total quality control or company-wide quality control, statistical quality control, training, quality control circles, and top management's understanding and support. These characteristics and some of the more important benefits gained by the Japanese firms which implemented them are: improved productivity, higher quality, motivated employees (white as well as blue collar), lower cost, and better competitive position, as shown in Fig. 1.

The double-headed arrows in Fig. 1 symbolise that these characteristics are interrelated. The existence of all these characteristics is essential for maximising the benefits of the Japanese quality control systems. It is possible to have just one or more characteristics exist in an organisation and achieve lesser results. An example is the case of the widespread use of quality control circles in the West, which have had little success compared to their achievement in Japanese firms.

Meeting customers' quality requirement

One of the most important aspects of quality control in Japan is the way that it is perceived and understood. The Japanese direct their efforts and activities in a way that satisfies customers' quality needs. In fact, meeting the consumer's demand for quality is the blueprint for various steps taken toward improving the quality of products or services. However, the final determination of whether quality has been achieved or not is reserved for the customers.

Although American managers would agree with the above statement, their approach to satisfying customers' quality requirements differs greatly from the Japanese approach. The perception that has dominated the USA since 1920 is that good quality is costly and, therefore, the firm may forego 'extra' improvement in product quality to avoid the added cost. As a result of such attitudes, the role of quality from the viewpoint of consumers has slipped from management's priorities. Thus, it is not surprising that a 1982 study of customer satisfaction indexes showed that among the top ten most satisfactory automobiles, six were Japanese, three were German, one was Swedish, and none was American. Lack of attention to the needs of customers can probably be identified as the biggest problem in the American quality control system.

It should be mentioned here that the word 'customers' encompasses not only end-users but all people involved in producing an item in the various stages of the operation. For example, workers in downstream processes are considered the customers of parts made by upstream workers. Any quality shortcoming from upstream workers will require extra work and adjustments from other workers down the production line.

To reduce and avoid such problems, appropriate process design is of utmost importance. Production-related ideas from workers in the line are fed back into the process design. Such a designing procedure for the production process, in turn, will help workers meet customers' quality demands more easily. In this manner, quality is built into the work process and every employee does his/her best in the step-by-step stages of the production process. This ensures quality of workmanship at all stages – from purchasing to sales and service after sales.

Statistical quality control (SQC)

Shewart's [2] first control chart at Western Electric during the 1920s introduced SQC. However, understanding Shewhart's work and its implementation required a good background in mathematics and statistics. Simplificaiton of Shewhart's work by Deming made it easier to teach SQC to employees in different levels in the organisation – from the top (management) to the bottom (production workers, janitors and so forth).

Different tools are used in the application of SQC. Table 1 shows several tools that are used in American and Japanese firms. The

Table 1   The most widely used statistical tools in American and Japanese firms

American firms*	Japanese firms†
Probability distributions and statistical parameters	Pareto charts
	Fishbone charts
Statistical inference	Stratification
Sample-size determination	Checklist
Control charts	Histograms
Confidence limits	Scatter charts
Accepting sampling plans	Control charts
Design of experiments	
Analysis of variance	
Scatter charts	
Basic reliability formulae	
Reliability sampling plans	

Note

American firms: SQC tools are used mainly by quality control department staff; only a few people in other departments may know how to use them.

Japanese firms: SQC tools are used by everybody, especially production workers. (Reduction in the number of quality control staff is possible.)

*Source: Quality Control Handbook, *Juran, J.M.* (Ed.). McGraw-Hill, New York, 2nd edition, Chap. 13, p. 3, 1962.

†Source: Ishikawa, K., Guide to Quality Control. *Asian Productivity Organization, Tokyo, 2nd revised edition, 1982.

important point is that among tools used in the American firms, such as statistical inference, analysis of variance, design of experiments, and reliability sampling, many are known and used only by an elite group, i.e. quality control engineers and staff both in Japanese and American firms. However, Japanese employees do know about SQC charts and those SQC tools which are relatively easy to understand and to use.

Effective implementation of SQC was made possible through modification and adaptation of the techniques by Japanese organisations. The creation of several useful tools, shown in Table 1, was the result of the refinement of SQC. These tools have been, and are, of great value in Japan. Their effectiveness has been proven by outstanding quality and productivity performance by Japanese firms in the past two decades.

Although SQC remains in great use in Japan, it cannot achieve the long-term goal of a manufacturing system such as Just-in-time (JIT) production, which requires minimum or zero inventories at all stages of production. Since SQC provides standards for an acceptable quality level, it would not be the most proper means for accomplishing the ambitious goals of JIT. Therefore, Japanese firms have searched for new ways and developed new means to ensure zero defects and high quality parts. One of the means which has been developed is total quality control or company-wide quality control (TQC/CWQC).

Total quality control or company-wide quality control

During the 1950s, A.V. Feigenbaum [3] introduced the idea of total quality control (TQC) or company-wide quality control (CWQC). He believes there should be an integrated system to develop, maintain, and improve quality and quality efforts at all levels of an organisation to maximise customers' satisfaction of product quality. The idea behind TQC/CWQC is that production is a responsibility which must be shared by all people in the organisation, especially the makers of the products. Therefore, the production department has the primary responsibility for the quality of the product, yet everyone is involved in quality improvement.

The reception of TQC/CWQC in the USA was less than gratifying; it received only lip service. Therefore, the idea of TQC/CWQC was not implemented on any significant scale in American manufacturing firms. It is important to mention that, contrary to practices in American manufacturing firms, TQC/CWQC starts at the top in Japanese firms. The important facts[4] about TQC/CWQC are:

- Quality control involves all functions that deal with the product.
- TQC/CWQC involves everyone's job: from the secretary who does the typing and makes typing errors; to the salesman who does not properly promote the product; to the president who makes poor decisions. In other words, all employees in all jobs are vital components of the TQC/CWQC system.

- Recognising that the responsibility for quality should be shared by all employees.
- Top management must understand and support the TQC/CWQC concept. In addition, a strong commitment to quality and leadership for successful implementation of TQC/CWQC is required.

The incorporation of several basic principles into the Japanese TQC/CWQC concept has assisted Japanese management to fulfil their commitment to proper implementation of TQC/CWQC. The main principles used in TQC/CWQC are: easy-to-see quality, total process control, insistence on compliance, line-stop authority, elimination of rework teams, 100% check, and project-by-project improvement.

The Japanese are using some supporting concepts to fully implement these principles. Some of the more important concepts are: use of a quality control staff as facilitators, use of small lot sizes, less-than-full-capacity scheduling, total preventive maintenance and good house-keeping. These principles and other concepts have facilitated the widespread use of TQC/CWQC in Japanese manufacturing firms.

Training

Although many US manufacturing firms consider training of great value, there has been a lack of understanding of the real importance of 'quality' training. The trend in American firms has not favoured the long-term education/training of workers for quality control. As a matter of fact, the strict bureaucratic structure of American manufacturing firms and the obsession with cost minimisation rather than quality have led management away from the most important resource of the firm: the worker. Most manufacturing firms in the US plan for a short period of time with acceptable performance goals judged on profits generated. Long-run investment in employee training and skill development, which reduce output in the short-run, therefore contradicts with the predominant goal of profit-making in many US firms.

On the other hand, Japanese managers perceive training/education of their employees as one of the most important components of their system. Thus, they approach training/education programmes from a different perspective. J.M. Juran [5] believes that one of the three most radical changes that Japan made was the creation of a massive training programme. They trained hundreds of thousands of upper and lower level managers, supervisors, foremen, and millions of non-supervisory employees from 1950 to 1970. This massive training programme resulted in a workforce that was, and is, well-trained for improving quality and eliminating quality control problems.

Employees from all levels of the organisation are put under continuous training, thus, massive training is an ongoing process. Learning new techniques, improving problem solving skills, seeking new information about their work, and looking for other training programmes are a regular part of each employee's job. This training, which is both on-the-job (OTJ) and in classrooms, continues until retirement. A Japanese employee in a progressive organisation receives an average of 50 days of both on-the-job and classroom training per year.

Workers with better training and education are more efficient and produce higher quality goods. Thus, it should not come as a surprise that the Japanese produce higher quality products more efficiently. To achieve a successful massive training programme, the Japanese have used different approaches borrowed from the West, to which they added their own methods. One of the methods that the Japanese use is nationwide quality control promotion activities, which are organised by JUSE. This method includes conferences and various presentations about quality. Nationwide quality promotion programmes plus the concept of quality control circles are among the most important means for promoting quality and the control of quality. In fact, these programmes helped Japanese workers develop the habit of improvement, which is a rare commodity in firms outside of Japan.

Quality control circles

The quality control circle (QCC) programme is one method used by Japanese firms to enhance employees' work knowledge. Scholars such as K. Ishikawa [6], H. Kume [7], and W.E. Deming [1] consider QCC one of the major components of the Japanese quality control system. A QCC programme is organised when a group of employees, performing similar work in the same area, meet regularly, on a voluntary basis, to discuss common problems of their work area. QCC involves both management and workers. However, the majority of people directly involved in QCC are workers with similar jobs.

The objectives of a QCC programme are to meet the organisation's goals of increasing productivity and quality through extensive training/educational programmes and to increase employee participation and job satisfaction. However, there are several important points that should be noted about the nature of QCC programmes [8]:

- QCC is a voluntary group (grass roots movement) and its members are not installed by management.
- QCC members define the scope of projects and activities.
- QCC detects defects rather than preventing them.
- QCC reduces defects rather than eliminating them.
- QCC group life can be continuous (project after project).

A close observation of QCC shows that this programme is similar to some programmes in the West. As a matter of fact, the QCC is almost a recombination of a variety of features of work improvement programmes that exist in Western industry. Several types of motivational and work improvement programmes are used in American firms which resemble QCC programmes. Some of the better known of these programmes are: zero defect programmes, employee suggestion programmes, and work simplification programmes.

Availability of these work improvement programmes did not prevent American management from looking into the QCC concept, and some found it more attractive and promising than the programmes already existing in American firms. Reports show that about 1000 companies are involved and active in QCC programmes. By implementing QCC, American firms hope to improve quality, productivity, and human relations in their organisations.

The overall effectiveness and contribution of QCC in the area of quality control have not been as successful in the USA as they have been in Japan. Lack of preparation, lack of support from both unions and management, selection of a weak group leader, and loss of interest on the part of QCC members are among the most frequent reasons for failure of QCC programmes in the USA. Results of a survey of American companies which implemented the QCC concept show that only 28% of the QCC programmes were deemed satisfactory and successful.

In addition to increasing quality improvement skills of workers, successful QCC programmes have other benefits[9]:

- Improvement of the worker's ability to communicate efficiently.
- Creation of a more cheerful work environment.
- Increase in workers' consciousness of their responsibility and in relating their jobs to the goals of the organisation.
- Increase in workers' ability to concentrate on finding problems and solving them.

Although the benefits of QCC programmes may seem trivial to some firms, it is important to note that Japanese management is aware of the fact that QCC is not a panacea and is only one tool among the several which have helped their firms achieve higher quality and productivity.

Management understanding and support

If one attempts to pinpoint the most important characteristic of Japanese quality control, it would no doubt be Japanese management's support and commitment to quality and quality improvement programmes. Japanese managers have shown their support and commitment to quality control from the beginning of their orientation to the modern quality control concept. When JUSE invited Japanese management to attend the first formal lecture on quality and the control of quality, 'they all came'. In fact, many believed that the real reason for Japanese superior achievement was 'super-managers'.

One of the more important responsibilities of top management is accomplished by establishing basic quality policies and a well-organised quality system. Many American firms lack a clearly stated quality strategy because quality improvement programmes do not have an immediate pay-off. In many short-term oriented American companies, management will not support a quality policy which does not offer a quick cost-reduction opportunity. The approach of some American firms has been to avoid any investment in quality programmes, or any projects for that matter, which will show profit only in the long-run in favour of speculation to make money in the short-run.

One of the main criteria for measuring management performance in the USA is the bottom-line result, the return on investment (ROI). Investments in quality improvement are ignored or sacrificed because management is bound to show quarterly profit improvements, an inescapable worry in the American manager's mind. Unfortunately, many American top managers do not realise that better quality leads to better ROI. Results of a recent study reveal that improvement of quality in the long-run not only increases market share, it also leads to a higher ROI.

In most Japanese firms, on the other hand, ROI is not the main index or criterion for measuring management's performance. If a project does not show a profit in the short-run, management will not be replaced. Therefore, a Japanese manager can easily give top priority to a quality improvement programme and support it all the way. Deming [1] argues that Japanese management support is one of the more important elements in Japanese achievement in quality control.

Japanese management introduced and implemented numerous, successful programmes in developing their employees' potential. They willingly shared information and power with workers throughout the organisation. To show their support, Japanese top management generally attempts to eliminate any barriers that stand in the way of meeting quality goals.

Japanese management has proved that quality is a vital management function which must involve everybody. The tremendous effort of Japanese management, and its support of quality control since the early 1950s, has been the main reason for the successful transition toward a routine awareness of quality among Japanese workers. It is only through the efforts of Japanese top management that Japanese workers have acquired the habit of quality improvement.

Implications for American manufacturing firms

The analysis and comparison of Japanese and American quality control systems leads to some implications which may be helpful in enhancing productivity and improving quality by American manufacturing firms.

Management's quality leadership

Perhaps the most important implication is that management must put an emphasis on quality. Indeed, quality should be a management philosophy which is translated into a basic strategy giving quality the highest priority. In this sense, top management will strive to meet customers' quality requirements, rather than merely conforming to specifications, which is a narrow view of quality.

Top management in American manufacturing firms should assume the responsibility for quality leadership rather than assigning leadership to the quality control manager. Top management's quality leadership must be continuous and its commitment and support to improving quality should be never-ending. However, these changes in management's attitude about quality cannot happen overnight. Top management should undertake a training programme to enhance quality consciousness and prepare itself for promoting quality at all levels of the organisation.

Organisational culture

One approach to prove management's support for quality is to develop an organisational culture which fosters employees' creativity, creates pride in quality workmanship, shows the role importance of employees, and creates trust between employees and departments. Many studies show that the best-managed American firms have, more or less, the same beliefs and cultural characteristics as do the best-managed Japanese firms.

According to T.E. Deal and A.A. Kennedy[10], some of the characteristics of American well-run firms are: no formal organisational charts; few formal rules, meetings, and memos; flexible jobs; a system of self-correction; minimum status symbols; two-way communication; a carefully defined employee selection process; and well-structured, social organisational programmes. These characteristics are common in both well-run American and Japanese firms. The culture is created by taking several important steps. These include:

- Devoting more time for selecting new employees.
- Focusing on long-term versus short-term planning.
- Encouraging a consultative decision-making process.
- Focusing on continuous training programme.
- Creating group loyalty.
- Creating group-consciousness.

Development of an organisation culture will ultimately enhance quality of the product and the working life of employees.

Manufacturing focus

Another important implication is a change of the manufacturing focus from a volume orientation to a quality orientation. For example, instead of having a production system which encourages excess buffers and allows quality problems to hide in work-in-process, a production system such as a Just-in-time (JIT) manufacturing system could be implemented. The Japanese JIT system has no buffer which leads to a prompt detection of any defect.

Plant reconfiguration

A quality-oriented manufacturing system requires continuous control and inspection of quality. Continuous control of quality not only implies that everybody is responsible for quality, but also suggests that quality must be built into the production process. The plant's configuration and layout play an important role in setting up a production process with built-in quality. Plant layout should allow for two-way communication with fast and accurate feedback and a smooth flow of information.

Quality control department

The quality control staff's function will change from that of being a police force waiting to catch a defective product to that of being a facilitator and teacher of quality control methods.

The quality control staff trains and educates employees in SQC techniques and arms them with different problem-solving techniques. Therefore, these trained workers are able to inspect for quality and have the ability to solve quality problems. The workers' involvement in quality control results in a significant reduction of defective items and reduction in the number of quality control specialists.

Prevention-oriented manufacturing

Prevention-oriented manufacturing versus inspection-oriented manufacturing is another important point to be considered by management in American manufacturing firms. A major step toward the improvement of quality is the use of a prevention-oriented manufacturing concept to improve quality of design. In the design stage, the relative cost of eliminating quality problems is low as compared to later stages in the production process. Thus, it is imperative to emphasise improving the quality of design.

Quality of design

Technological innovations such as computer aided design (CAD) and computer aided manufacturing (CAM) are playing an important role in improving the design. CAD/CAM reduces time lag for any changes in design; therefore, adjusting to changes in the product design required by customers will be done quickly. However, the promised advantages of CAD/CAM will be realised only if there is good communication between the production and design departments. In fact, some of the best managed firms in the USA are merging their design and manufacturing departments into a continuous process which allows for the smooth flow of information.

Vendor management

Incoming raw materials establish another important element in improving quality. Undetected defects in raw materials will create problems somewhere down the line. The further the defect is carried, the more costly it will be. Detecting the defective item at the source (supplier) should have the highest priority for management. Vendor selection then becomes one of the most important steps in providing 100% defect-free, raw materials. The other important factors in improving quality of incoming raw materials are supplier evaluation, shipping and receiving, product specification, lot size, and transportation. In analysing all these elements, American management's emphasis should be on quality rather than on price. Such a vendor management system will minimise defects and improve quality.

Concluding remarks

Goods-producing industries represent the backbone of the economy of industrialised nations. Although the growth rate of service industries has been astonishing (it accounts for 70% of the workforce in the USA), this growth depends on the existence of a strong and healthy manufacturing sector. The Japanese have shown that improvement in the performance of manufacturing firms can come from producing higher quality products.

Comparing the Japanese quality control system with the American quality control system shows that quality should be treated as a management philosophy rather than a sub-goal which is handed down to middle management. The Japanese have proved that improving product quality is not as simple as using a set of statistical formulae; rather it is the combination of statistical methods, hard working employees, advanced technologies, and most of all the management philosophy which has its heart set on serving the customer. However, putting all these elements together may solve only a small portion of the quality problems. The key to management's role is in creating the organisational culture which fosters employees' creativity for quality.

Some implications for American management have been suggested to improve quality performance in manufacturing firms. These recommendations may encompass all functions at all levels of manufacturing firms. However, it should be pointed out here that these implications may require fundamental changes in the organisation of manufacturing firms and may require modification or creation of a proper organisational culture.

References

[1]   Deming, W.E. 1982. *Quality, Productivity, and Competitive Position*, Massachusetts Institute of Technology, Center for Advanced Engineering Study.

[2] Shewhart, W.A. 1931. *Economic Control of Quality of Manufactured Product.* D. Van Nostrand, New York.
[3] Feigenbaum, A.V. 1961. *Total Quality Control: Engineering and Management,* McGraw-Hill, New York.
[4] Rilker, W.S. 1983. QC circles and company-wide quality control. *Quality Progress,* 16(10).
[5] Juran, J.M.(Ed.) 1982. *Quality Control Handbook,* 2nd edition. McGraw-Hill, New York.
[6] Ishikawa, K. 1984. Quality standardisation: Program for economic success. *Quality Progress,* 17 (1): 16-20.
[7] Kume, H. 1980. Quality control in Japan's industries. *The Wheel Extended,* 9 (4):20-27.
[8] Loeser, E.A. 1983. Management commitment to quality: Rockwell International Corp. *Quality Progress,* 16(8).
[9] Juran, J.M. 1967. The QC circles phenomenon. *Industrial Quality Control,* 23(7).
[10] Deal, T.E., and Kennedy, A.A. 1982. *Corporate Cultures,* Addison-Wesley, Reading, Massachusetts.

CONTINUOUS IMPROVEMENT THROUGH STANDARDISATION

R.W. Hall
Indiana University, USA

The correct use of standards can lead to great improvements in manufacturing performance. However, it is important to understand related disciplines and activities. When properly used, standards are indeed tools of progress and not constraints on creativity. If standards include an explanation of why certain steps are taken they are much more likely to be adhered to and become a *practice*, not just a procedure. Making operators responsible for the accuracy of instructions is also important in achieving consistency. Today's standard is the foundation for tomorrow's improvement. Indeed Henry Ford's ideas on continuous improvement are the root of the development of Just-in-time production.

Standardisation has been important to industrial progress since the dawn of mass production, and it will continue to be so. Yet standardisation is a widely misinterpreted function of manufacturing. To gain the greatest manufacturing performance improvement leverage, we need an understanding of related discipline and activities.

'Standardisation' evokes earlier images. For many manufacturing executives, the image is not an exciting one. These executives may have started by writing industrial engineering standards in one form or another – those dull, detailed listings of machine, method, material, and time, nested in archival tomes somewhere in a production office. These weighty writings can be referenced whenever someone needs to get to the bottom of 'how it really should be done'. When referenced at all, such standards generally are used to make a point in an argument – which may or may not settle the argument.

Personal experience with other 'standards' can be just as tiresome. The term is applied very broadly:

- *Government standards* – regulations and codes.
- *Industry standards* – fasteners, wire gauges, containers, barcodes, and what seems like an infinite list of things covered.

- *Company standards* – nearly every aspect of life is referenced: materials, training, costs, times, and so forth. Bills of material and lead times are standards essential to both engineering systems and materials systems, for instance, and product specifications are standards for both engineering and quality.

Although they are essential to the fabric of corporate communication, standards are not regarded as having much pizzazz. The term 'standardisation' seems to connote repressed creativity – standards prevent us from doing everything in any way we choose. That is their purpose. Not all deviations are improvements. Properly used, standards are tools of progress.

Choosing the best method

Difficulties in understanding the use of standardisation are not new. Henry Ford discussed this situation nearly 60 years ago[1]:

"To standardize a method is to choose out of many methods the best one and use it. Standardization means nothing unless it means standardizing upward."

"What is the best way to do a thing? It is the sum of all the good ways we have discovered up to the present. It therefore becomes the standard. To decree that today's standard shall be tomorrow's is to exceed our power and authority. Such a degree cannot stand. We see all around us yesterday's standards, but no one mistakes them for today's. Today's best, which superseded yesterday's, will be superseded by tomorrow's best. That is a fact which theorists overlook. They assume that a standard is a steel mold by which it is expected to shape and confine all effort for an indefinite time. If that were possible, we should today be using the standards of one hundred years ago, for certainly there was then no lack of resistance to adopting what goes to make up the present standards."

"Industry today, under the impulse of engineering ability and engineering conscience, is rapidly improving the standards. Today's standardization, instead of being a barricade against improvement, is the necessary foundation on which tomorrow's improvement will be based."

"If you think of 'standardization' as the best that you know today, but which is to be improved tomorrow – you get somewhere. But if you think of standards as confining, then progress stops."

Ford was a supporter of continuous improvement, and many of his ideas are the root of the development of Just-in-time production. However, the relationship of standards to this improvement process is not heavily publicised.

Consistency and quality

Standards promote consistency, and consistency is one of the major foundations of good quality. *Use* of a standard assures that once a better method has been found, all in an organisation will use it until something better supersedes it. However, merely issuing a standard in writing does not do this. Discipline in the development and use of standards does, and that is part of progress.

For a lesson in the value of a good standardisation in practice, visit a McDonald's restaurant anywhere in the world. Compare one of their outlets to a franchise that 'can't quite get its act together'. Personal appeal of a Big Mac is not the issue. Consistency is, and a Big Mac is a Big Mac is a Big Mac – anywhere – like it or not.

Consistency is a major factor in quality performance and quality production. How do you know if you have effective standardisation? Watch a multi-shift operation, such as a plastic moulding operation, at shift change. If the first action of a number of operators is to begin tweaking the set-up conditions of the previous shift, standardisation has a way to go.

Why do the operators do that? Because no one has a fixed idea what the proper operating conditions of each part on each press should be. The quality specifications for acceptable parts also may be slightly 'adjustable'. Everybody then has a different idea about it in practice, although the specifications and standard sheets may have been around long enough to have turned yellow. The reasons why operators alter running conditions vary. They may be attempting to touch up the quality of the part, or make the machine run a little better, or just 'lay it in a groove' so they do not need to check it for a while as they research the point spreads for the weekend ball games.

Unravelling this situation and arriving at 'correct' operating conditions takes work picking through many variables. Specifications, mould, mainteneance, material – all should be free enough of variance that every operator can set the equipment to the same settings and agree that those are the best. Standardisation is a disciplined company-wide *practice* as well as a procedure. Operators need not only be aware that a standard exists, but must understand it and agree with it. Otherwise, they will not follow it.

Practice is the backbone of detailed improvement. Without this discipline, improvements once attained cannot be held.

Ryuji Fukuda[2] emphasises this strongly in his discussion of quality improvement, particularly in conjunction with his CEDAC method (Cause and Effect Diagram with the Addition of Cards). One of the primary problems the method is intended to address is the differences in perception of operations by different people, just as in the case of the plastic moulding operators. The CEDAC approach is powerful if a group applies it with diligence; but doing that consistently is a problem in Japan, as it might be elsewhere, and two of his exhibits bring that out.

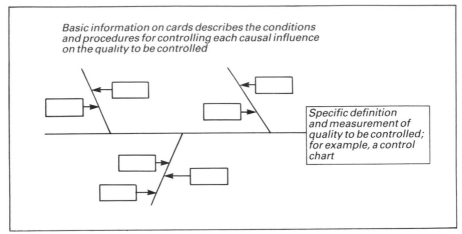

Fig. 1   Basic CEDAC

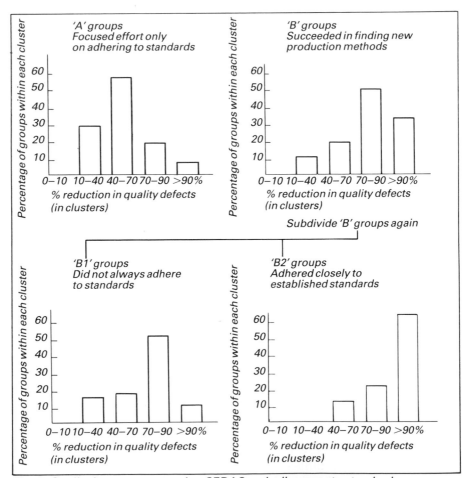

Fig. 2   Quality improvement using CEDAC and adherence to standards

CEDAC (Figs. 1 and 2) is a very basic addition of a communications method to the Ishikawa cause and effect diagram. Many variations are possible. The whole idea is but one approach to a large subject area – design of experiments, but one which recognises one of the major problems of experimentation involving many people – consistency of direction through discipline and communication. How many engineers have wondered in the morning what an operator did on the night shift?

As Fukuda points out, adherence to a standard is an important element in making progress. If you do not know where you are, you cannot be sure where you are going. That is a very basic concept. But basic is not easy.

The conclusion if Fig. 2 is that adherence to standard is very important to progress. Even the 'A' groups which mostly emphasised adherence, not experimentation for improvement, made good progress. This is because once someone else made an improvement, they followed the new standard. The biggest gains came from the B(2) groups which did both. One of the problems of improvement is holding gains once made.

Visibility and operator responsibility

When touring a top Just-in-time plant, one factor to look for is visibility of operator standards and directions. Instructions for anything important should be out where operators can see them, not filed away and forgotten.

Another important practice is to make operators (and supervisors) responsible for the correctness and currency of their instructions. They may not write them in the original, but they use them and interpret them. Once they know that their work methods are correct, the instructions should describe them clearly enough that another experienced operator can use the instructions to perform the same job in a correct manner. Rotate workers in their positions from time to time to try this out.

In many plants the 'working standards' are kept in people's heads. Changes in procedure are communicated, but only one or two persons understand them thoroughly. If they happen to be absent or are transferred, regression in procedure occurs. The evidence of this is found in dog-eared, dirty, little notes, such as 'Check with Bill about this'. If 'Bill' is not convenient to locate, the operator will figure it out as best he can. Unresolved quality problems, or other problems, lead to more headaches. As a result, standardisation is not effective.

Writing a standard is an exercise in communicating detail. A standard should not be a legal document useful for fixing blame after errors are made, but a working tool communicating to everyone necessary how to perform properly, perhaps even with an explanation of 'why' thrown in. This process is one of the major reasons for face-to-face communication between staff (engineers) and hands-on personnel.

Writing standards

Experience in doing this job well is not developed in a few days – the time frame is more like a few years. It does not help if writing standards is considered by many to be 'grub' work of little consequence. (Both Ford and Ohno considered writing a standard of this kind to be among their most valuable educational experiences.) The details are the bedrock of manufacturing improvement.

In the Toyota system a standard consists of three parts:

- *Cycle time:* The time allowed for performing one cycle of making the part derived from the cycle time of use for the part (inverse of the frequency of use of the part at assembly). If the cycle time of use substantially changes, the method for making it may substantially change, a change in fabrication procedures roughly analogous to rebalancing an assembly line. The objective is to come close to making parts at the rate they are needed.

- *Sequence and detail of work:* Specifics of how to do it, and how to do it correctly. How to accomplish each of the specifications required, failsafe methodology to be employed, and so forth.

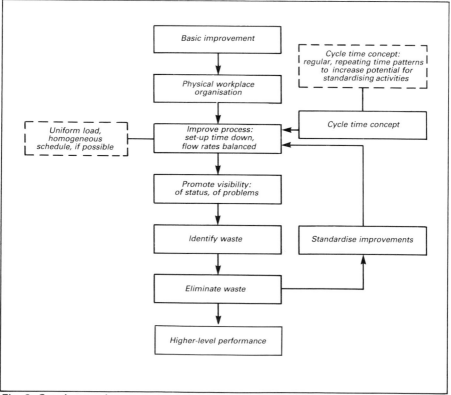

Fig. 3 Continuous improvement through standardisation

- *The standard WIP:* This is related to the amount of time allowed to recover from a problem should one occur, allow for possible changes in mix of parts required, and so forth.

Standards are also written for set-ups in order to perform them quickly. These standards are very important to quality. The first piece made from a subsequent set-up should be identical to the last one made from the prior set-up. If machine, tooling and procedure are developed to this point, any extra tool wear from frequent set-ups should disappear, and the checking of quality at set-up times should add to the assurance of it. Well-done set-up reduction is a process very much like any other experimentation to be performed in production, subject to the methods of standardisation.

Fig. 3 illustrates the way in which standardisation fits into the overall approach to 'Just-in-time manufacturing' in the repetitive case for companies which have the potential for it. However, standardisation is still important for companies whose potential for repetitive work is somewhat limited.

Done well throughout an organisation, standardisation should clarify everyone's job, if not make it easier. Improvements should be possible to prove, and one person's improvement not undone by others. Standardisation affects every aspect of manufacturing. It is part of the glue holding things together. Some examples:

- Drawing methods.
- Bills of material.
- Engineering change methods.
- Tooling development.
- Specifications.
- Gauging, test methods, test equipment.
- Maintenance procedures.
- Programming documentation.
- Order entry.
- Work-to-schedule discipline.

How standards work

NOK Inc. is a Japanese-owned company, but now with many Americans in the management of the plant at LaGrange, Georgia. It is not Toyota. The company is developing itself to perform needed operations, as is true of any company, but the development of standardisation is an important part of their process. One can see the philosophy of effective standardisation at work.

At NOK, standardisation is used with at least four kinds of understanding, and perhaps more:

- Unification of concept – cohesive action. A single activity fulfills multiple objectives. Simplicity. For instance, much of the detail of their quality system is packed together into one unified chart.

- Documented, consistent procedures and conditions: quality, methods, maintenance, and so forth; *detail* is provided.
- Retaining one-time improvements in practice until better are found.
- A step-by-step methodology for questioning old assumptions and developing new improvements. In brief, this methodology is: seek causes (ask why five times, etc.). Solve. Check solution. Standardise. Do it again.

The perception by each person of job scope and responsibility is strongly affected by this methodology. Production control has a major responsibility for quality, and quality control is essential to stay on schedule. There is no strong separation of functions. For instance, a single tag in the standard container serves to gather information for both quality and for materials control. They think of one system, not many, and the system design comes from the mental set and not the other way around.

NOK is doing well, but they have not gone as far as they can go. Their case illustrates the construction of a very powerful manufacturing system, brick by brick.

References

[1] Ford, H. (with S.Crowther). *Today and Tomorrow*. Doubleday, Page & Co., New York, pp. 80–81.
[2] Fukuda, R. 1983. *Managerial Engineering*. Productivity Inc., Stamford, Connecticut.

A FRAMEWORK FOR JIT IMPLEMENTATION

J.R. Bicheno
University of the Witwatersrand, Republic of South Africa

The concept of Just-in-time manufacturing is explained by means of a two-stage framework. The actions in Stage One prepare the plant, processes and products for more efficient production, and are applicable in every manufacturing concern. The Stage Two actions execute Just-in-time but success is more dependent on company culture. Together, the two stages form an ongoing cycle of improvement, which may be seen as the integrating framework for techniques such as materials resources planning (MRP II) and technologies such as FMS.

It is perhaps unfortunate that the phrase 'just-in-time', or the acronym 'JIT' has entered the manufacturing vocabulary. It is unfortunate not because Just-in-time is most descriptive of what the aim is, but because far too many manufacturing managers and an even greater number of general managers regard the phrase with suspicion. We live in a world of catch phrases and buzzwords. Unfortunately many of these phrases are presented for marketing purposes as breakthroughs or as 'quick fix'. Naturally enough, therefore, all such phrases or acronyms are regarded with equal scepticism.

In fact, Just-in-time has evolved into a set of concepts as to how operations management should be conceived. Operations management is more general than manufacturing management, both within a particular manufacturing organisation and from the point of view that it is applicable to organisations outside of manufacturing. Just-in-time has evolved from being seen as an approach to scheduling (to be compared with, say, materials requirements planning [MRP]) to a set of concepts which view scheduling in a more fundamental way (including, for example, set-up reduction and layout), and now as a philosophy which guides every aspect of operations management. Just-in-time has emerged as an approach which outdates very large sections of operations management theory. It is a watershed.

For the manufacturing manager perhaps the greatest benefit of JIT is that it provides a new conceptual framework for operations management. The aim here is to present one such framework based on JIT concepts, and to show how this can be used as a guide to implementation.

It is interesting to note that prior to JIT there have been few conceptual frameworks which provide practical guidelines or a path of action for operations managers. An examination of pre-JIT textbooks will confirm this. Typically, topics such as layout, quality, maintenance, and scheduling were presented independently with minimal cross-reference. Perhaps this has encouraged the practice, so often seen in Western factories, of tackling apparent problems independently and then being disappointed with the results.

The JIT goal

JIT has a provocative goal which may be stated as: 'to produce instantaneously, with perfect quality and minimum waste'. The goal will never be achieved, but the direction is clear and the journey can begin.

The word 'instantaneously' requires some explanation. The ideal way to produce the end-product is literally just in time to meet the market demand for it. Thus, JIT is primarily a lead-time reduction programme. But, this form of last-minute production will only be the most profitable and wasteless way if we have fairly uniform demand. If the demand is not uniform then either we need to strive to make it more uniform (is the demand 500 per quarter or is it really one per hour ?), or we need to become sufficiently flexible so that production as late as possible becomes closely correlated with profitability. 'Instantaneously' relates to lead-time, and hence to bottleneck operations. It is necessary therefore to first tackle problems in bottleneck operations to facilitate the 'keep it moving' or minimum lead-time aim. 'Waste', of course, refers to any activity including movement, storage, delay, or even inspection, which does not add value to the product. However, one would need to consider very carefully whether 'waste' in non-bottleneck operations is in fact waste at all. (For example, improving the cycle time of an under-utilised machine.)

There is another implication of the concept of meeting market rates of demand 'instantaneously'. That is that a single organisation will rarely be involved. The complete chain of organisations, from raw material producer to final seller, must all ultimately be synchronised without any intermediate waste of time or resources. If they can be, the JIT philosophy is that cost will be minimised and competitiveness and profitability maximised. This is the ultimate challenge.

The JIT framework

The framework for approaching the goal may be presented as a two stage process:

- *Stage One* is concerned with preparing the facility and the product for: high quality, low cost, minimum lead-time, and high flexibility. In other words, Stage One sets the essential structure in which the subsequent processes of approaching the JIT goal may take place.

- *Stage Two* undertakes those processes which are necessary to produce: instantaneously, to market rates, with perfect quality, and minimum waste.

There are various activities that need to take place in each of these stages. These are shown in Figs. 1 and 2 and will be used to discuss in some detail how JIT 'implementation' may proceed. At this point, however, it should be noted that:

- The activities are, for most manufacturing and operations facilities, the minimum necessary to approach the ultimate goal. This is important. It means there is no 'quick fix'. Operations competitiveness cannot be achieved by, for example, putting in an MRPII system or an FMS. These are at best only partial solutions, and at worst are dangerous extravagances unless seen as part of the wider JIT framework. What is needed is a coordinated step-by-step action on a broad front.

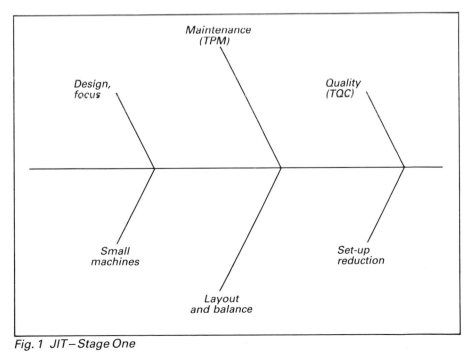

Fig. 1 JIT–Stage One

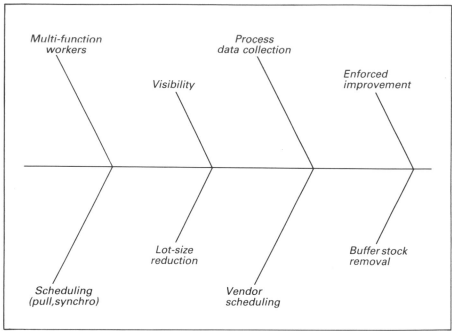

Fig. 2   JIT – Stage Two

- The stages are not strictly sequential. The Stage One activities will very often be the best way to move into JIT (or indeed to stay in business) but not in all cases. Although virtually every activity mentioned in the framework will have some applicability in every operations organisation, the emphasis will differ widely. More usually, the stages will form a cycle of improvement. Making progress with the Stage One activities will facilitate Stage Two activities. Likewise, making progress with Stage Two activities should require further Stage One action, and so on. There are many possible paths. One path may be improved maintenance leading to reduced buffer stocks leading to improved layout (reduced flowpath) leading to better visibility leading to improved quality, and so on. To reflect the ongoing cycle, Figs. 1 and 2 are presented in the form of Ishikawa or CEDAC[1] cause and effect diagrams (Fig. 3). One might visualise improvements being built upon previous experience, but always contributing towards the required goals.

- 'Implementation' under JIT is perhaps a misleading term. JIT is not implemented as perhaps an island of automation may be. With reference to the goal, JIT will never be totally implemented. JIT implementation is an ongoing process.

- Hence, in undertaking the cycle of JIT improvement, it must be understood that JIT *is a strategy*. By definition, a strategy contains the following elements[2]. It involves:

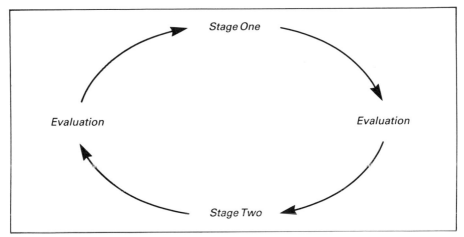

Fig. 3 Cause and effect diagram

- an extended time horizon. It is a guide not only as to what to do next, but also as to what will still be the guiding philosophy next year.
- significant impact on the 'bottom line'. The various activities must be undertaken to directly influence the bottom line, not because it merely seems a good thing to try. Effectiveness against the goal must be demonstrated.
- concentration of effort on the minimum necessary activities. This implies not being side-tracked into peripheral activities which demand inordinate management time (perhaps a new manufacturing information system which addresses only a small part of the total picture).
- a consistent pattern of decisions. For example, new products or machines which do not fit in with the overall JIT goal, or which would steer the organisation away from the JIT path would merit particularly careful consideration.
- pervasiveness throughout the organisation. JIT very definitely has implications for design, marketing, accounting, and personnel. It is not the sole concern of operations. This realisation may be the single greatest barrier against achieving the full potential.

The various JIT activities are now briefly discussed, paying particular attention to the question of implementing JIT. For further details on the various JIT techniques the reader is referred to the excellent texts listed in the references at the end of the paper[1–7].

At this point it is worth emphasising the point that JIT is not an alternative to MRPII, FMS, or CIM. Rather, JIT is a strategy which may incorporate these and more. The Stage One activities are applicable whether or not one wishes to adopt MRPII as Stage Two activity. In fact, MRPII is likely to work much more effectively if the Stage One activities have been addressed.

Stage One

Design and focus

This is concerned with rationalising the product line, and then engineering and simplifying the chosen products for ease of manufacture. This aim follows the Japanese approach of selecting a few products with wide appeal, rather than a wide range of products each with selected appeal. In this way volume and the associated learning curve effect is maximised.

Focus is related to the concept of the 'focused factory' as first proposed by Wickham Skinner of the Harvard Business School. This concerns identifying those few actions and characteristics that are crucial to success, and doing these very well even at the expense of others. For JIT this would imply thinking in terms of 'factories within a factory', each with its own specific requirements and concentrated expertise. Contrast this with the mixed factory trying to produce a range of products having different key success factors.

It is clear that 'design' and 'focus' are interrelated. Perhaps the most powerful way to move towards focused factories is through design. Under JIT, design cannot be a divorced activity, but must become integrated with manufacturing. This, of course, like many individual JIT principles, has been good management practice for many years. Perhaps what is new is to see such actions as contributing specifically towards the JIT goal. Current examples include deliberate organisational merging of design, manufacturing, engineering and quality, and even involving suppliers in product design; deliberate product line rationalisation; programmes to increase 'part commonality' within a product range; and programmes to reduce tooling variety. In the case of these last three, 'group technology' concepts in combination with CAD/CAM can be seen as a vital facilitating tool for JIT.

All these actions may be seen as leading towards the overall JIT goal. As with all the Stage One activities they are beneficial in their own right, show results on the bottom line, but are still synergistic with the JIT goal.

The discipline of 'value engineering', applied both to the product and to the manufacturing process itself, is often an essential part of the design and focus activity. The 'value' concepts are powerful in themselves but especially so if they are directed at the same overall JIT goal. Value engineering is in fact totally compatible with JIT especially with regard to the team approach to creative improvements.

With regard to implementation priorities, design and focus has to be a longer term, ongoing aim. Companies having a very diversified product range, especially if they operate in a make-to-order environment, may not be able to move significantly into Stage Two before they have achieved design and focus rationalisation. On the other hand, high volume or limited range manufacturers will often not require design actions, but focus concepts may still be highly relevant.

Total productive maintenance, total quality control, and set-up reduction

Just-in-time concepts have highlighted the necessity of bringing these three activities to the fore in any operations environment. They are sometimes considered to be the fundamental legs of a JIT programme.

From an implementation point of view the approach to all three is similar and should be undertaken in every operations environment. To start with, a process chart is developed for every major product. There will usually be an 'official' and an 'actual' chart, with the actual chart showing all informal and non-planned activities, such as rework loops, delays and temporary storage. Process charts should also be developed for maintenance and set-up activities, particularly those associated with bottleneck operations.

All these process charts must now be examined in detail and the individual activities classified. Classification takes place in two ways. First, all activities are examined from the point of view of whether they *add value*. For example, delays, storage, transport, inspections, and rework do not add value. All activities that do not add value are categorised as waste and are candidates for elimination. A second stage concerns the priority of waste elimination. Some waste categories can be eliminated in the short term, whilst others may require a long-term ongoing effort. Cost effectiveness must now be brought in to produce the priority list for waste elimination. However, the overriding priority is to reduce waste at bottleneck workcentres, so as to allow flow-type production.

Secondly, all activities that are not immediately eliminated as waste, including both value-adding and longer term waste categories, are considered from the point of view of *who* is actually to do it. The principle is that operators themselves should do as much as possible, rather than the specialist quality controllers, maintenance men, or set-up technicians. Over time, training must take place to allow the operators to take over more and more of the specialist tasks. Again, a priority list can be drawn up. This also has implications for the role of work study and quality control in the JIT plant. No longer are these the preserve of specialists, but become everyone's concern with former specialists taking over in the role of coach or facilitator.

In the case of set-up reduction, a further categorisation aspect must be made between 'internal' (those activities that can only be done whilst a machine is stopped) and 'external' (those preparatory activities that can be done before the actual set-up). Here, the idea is to do as much of the set-up as possible off-line, and in time to move as many activities as possible to the external. Note that set-up is not only machine changeover, but also the support activity of parts access.

Total quality control is so central to Just-in-time that it is perhaps inappropriate to list it as a Stage One activity. In fact, the majority of activities in the two stages are themselves important contributions in the journey to zero defects. If there is poor quality, and in particular if there are high variations in quality, then JIT is severely hampered. On the other hand, if there are JIT elements in place then TQC is made easier. For implementation this would appear to be a 'chicken and egg' situation, yet it is not. JIT and TQC must both be implemented together, not in parallel by two non-communicating teams. The problem is how to step onto the feedback loop. The essence of TQC is defect prevention by involving everyone from design and supply to final delivery. It is a people programme, well encapsulated by the Zulu word used at Toyota South Africa – Eyakho; meaning build it as if it were your own, understanding the reason why it is necessary. So, start with people moving towards understanding, responsibility and authority. However, if process quality control is not already in place, it must be a priority consideration. Often the swing from inspection-based quality to prevention-based quality is a lengthy one, so an early start is called for. The process categorisation exercise mentioned earlier should give an indication of priorities.

Layout and balance

Layout follows the steps taken in the previous section. Under JIT, layout is not a static concept, but rather a continuous drive towards shortening the manufacturing process length. As each category of waste is eliminated, the process length should be shortened – often, of course, it is the principal way to eliminate or reduce wasteful movement.

Layout also follows directly from 'focus' concepts. Here, the concepts of group technology (GT) are most relevant. GT seeks to group machines so as to produce families of products with similar manufacturing characteristics. Where families can be identified it should often be possible to regroup machines into cells which provide for the minimum of set-up time and of material handling without any inter-operation storage. The identification of possible families is therefore a crucial early step in JIT Implementation.

Housekeeping, as in all well-managed operations, is important. Under JIT, housekeeping becomes a structured task well described by the Toyota principles of 'Seiri' (putting away everything that is not needed for the chosen time bucket) and 'Seiton' (arranging those things that are needed in the best possible way). As with set-up, quality and maintenance, the ideal is to involve operators in this task to the maximum so as to allow ongoing improvement. Taking this concept further, JIT becomes the best possible way to move into automation since arrangement and parts location problems reveal the most cost-effective automation actions.

'Balance' is the concept which relates to uniform plant load and to manufacturing to market rates of demand. Clearly 'making a little each day' rather than a large batch periodically has fundamental implications for layout. Of course, balance cannot be achieved overnight but requires ongoing effort in design rationalisation, focus and marketing. Also, balance is made easier by using 'small machines'.

Small machines

Using the smallest machine possible, consistent with quality, is a long-term commitment by management to ensure flexibility. The small machine concept, where several small machines, possibly permanently set up, are preferred to one larger one, is consistent with the concept of producing to market rates of demand. One larger machine may simply make this concept uneconomic. In particular, older often smaller machines, permanently set up may be used at bottleneck workcentres.

The small machine concept extends to 'Nagara' (self-developed machines or low-cost automation) and to 'Pokayoke' (fool-proof devices which prevent defects). This is simply about in-house manufacturing innovation that allows improved quality, reduced waste, and dramatic shortening of lead-times by combining operations. Numerous examples now demonstrate the potential of lead-time reduction and quality improvement through in-house engineering of processes. In-house innovation in manufacturing technology is not easy to implement, particularly in organisations that have seen these skills as unnecessary overheads and cut them back. Although Nagara originally involved essentially mechanical engineering skills, the emphasis is increasingly on electromechanical and microcomputer skills. Although some skills can be purchased from outsiders, the essential opportunity recognition often cannot be. Therefore, for JIT implementation, a commitment towards in-house engineering skills should exist. Obviously, the small machine concept is a long-term process and it is vital that those making machine purchase decisions see the economics in terms of the full Stage One objective rather than narrow issues such as cutting speed or set-up taken alone. One purchase of a large fast machine which nevertheless produces a single bottleneck, vulnerable to breakdown, can delay JIT potential for a long time.

Stage Two

Stage Two takes the structure created in Stage One and develops the process of making only as needed, with minimum waste. We may note that Stage Two is 'detachable' – that is, organisations not yet willing to take the leap into full JIT may prefer to omit it. Stage One on the other hand is applicable to virtually every operations organisation.

Multi-functional workers

Like several JIT concepts, multi-functional workers exist in many non-JIT plants. The concept of having workers capable of performing several tasks, and perhaps basing their reward system on this capability, is consistent with the ability to make only what is needed only when it is needed. JIT demands that operators be available to be moved, perhaps even in the same shift, from low to high demand areas.

Again, implementation is a longer term task but with the ultimate aim being clear. Under JIT, the multi-functional worker concept takes on the added view that the ideal JIT operator is also responsible to a large degre for method improvements, and aspects of maintenance and quality. It is therefore inevitable that training in a JIT plant is likely to be far more comprehensive than in the non-JIT plant. This means that the JIT implementation plan must include early consultation with unions or worker representatives. It must be understood that management or engineers do not implement JIT, but almost everyone does. Perfect quality with minimum waste cannot be otherwise. The multi-functional worker concept applies as much to managers and staff as it does to operators. Indeed, it is these former groups that often resist the new roles most strongly.

Visibility, process data collection and enforced improvement

Visibility is concerned with making problems visible by suitable immediate displays on the shop-floor (not inside a computer!). Process data collection concerns ongoing, often automatic, updating of data relating to production rates, quality, maintenance and set-ups. These are related issues of management style under JIT. The aim is continuous, enforced improvement by making problems visible and demanding quick response to solve them. The philosophy is encapsulated by Toyota which permits ordinary operators to literally stop the production line rather than make defective products. Line stop is welcomed because it identifies a problem that can then be solved. But, and this is central, they do not expect the line to stop for the same problem more than once. Immediate and total response becomes the norm.

Although there is the overriding JIT philosophy of solving problems at the lowest possible level, JIT is also about 'living dangerously'. This means a willingness to operate with almost no margin for error, in the belief that only by being close to the limit will true problems not only be exposed but will demand attention. Waste must be discovered and reduced, and the prime tools are an ongoing aim to reduce buffer stocks and lot sizes. Bottlenecks should become more apparent when they are not protected by inventory. This makes the location for the next set of implementation priorities clear. This is interesting. It means that a JIT implementation sequence cannot be mapped out well in advance by some remote planner. Rather it must be worked out by 'hands-on' management. This is not fire fighting, but orderly, incremental progress towards the JIT goal. Again, implementation under JIT is ongoing. Following Stage One actions in set-up reduction and layout, an appropriate cut in lot sizes and buffers must follow.

Living dangerously in an atmosphere of forced-pace change does not suit all managers, particularly those used to managing by remote control through layers of hierarchy. JIT is about bringing small-company, hands-on-type management to all manufacturing companies. This is the real cultural change, rather than the more publicised Japanese cultural characteristics. Under JIT, operations become a prime competitive weapon – at least as important as marketing and finance – and so general management can be expected to be seen on the operations floor 'managing by walking about'.

Walking around a JIT plant is usually a lesson in graphic arts. Instead of the usual well-intentioned, but often meaningless, quality and safety slogans, one sees a host of graphs and other data prominently displayed and up to date. These would include Ishikawa or CEDAC diagrams, X-bar and R-charts, reject rates, mean time between failure data, standardised sequence steps for set-ups, set-up time history, multi-functional skill charts, aircraft-style maintenance checks, quality and maintenance photographs, Pareto problem flip charts, and more. Sound and vision have an important place – such as light systems to indicate quality, maintenance or delay problems, and sound to indicate sequence delays.

The visibility and process data collection concepts relate also back to layout and design. Quality specialists, maintenance crews, designers, and operations managers should all be grouped not only physically together, but as near to the shop-floor as possible in order to facilitate better communication and faster problem solving. In a South African factory designed in Japan several supervisors found they had no offices, simply because the Japanese designers assumed they would spend all their time on the shop floor!

The interdependence of JIT techniques is shown by the concept of under-capacity scheduling which sets aside time for problem solving each day or each week. The problems to be worked on, with operator involvement, are clear because light systems, line stop and visibility in general have highlighted those most pressing.

Scheduling, lot-size reduction and buffer stock removal

Scheduling under JIT implies producing to market rates of demand in the very short term, and in particular avoiding overproducing. The cardinal sin under JIT is to overproduce. In other words making anything, including components and subassemblies as well as final products that cannot be sold in the very short term must be avoided. The scheduling rule is to produce an ever-greater proportion of items just in time to be shipped. JIT tries to achieve continuous *flow* which minimises lead-time.

To implement JIT is therefore to move towards producing a set of products across the range, with quantities in proportion to the current sales mix, in a shorter and shorter time bucket. Scheduling aims at sending through a set of products which will meet the market demand for that time bucket, whilst not overproducing. Ultimately, as the lot size comes down, a complete market-paced range would be produced perhaps every week or even every day, in a high-volume plant. Regular schedules, comprising the complete proportional range, would be repeated with increasing frequency, but decreased number of items. Regularity with no surprises results. The prime requirement for a JIT scheduler is to know where the bottlenecks are and what their throughput rates are. The aim is to keep bottlenecks paced to market rates of demand in the short term which should mean high machine utilisation and operators occasionally idle. On the other hand, at non-bottlenecks one should find idle machines but perhaps fully occupied operators tending several machines.

The 'pull system' is an important JIT tool for scheduling. There are, of course, many variations and the reader is referred to the reference list. When the pull system is combined with regular batch sizes it becomes a prime tool to control inventory levels and prevent overproducing, as well as to identify bottlenecks. Knowledge of bottleneck capacities tell the scheduler about the maximum throughput rate, and the pull system automatically prevents overproduction at non-bottlenecks. Where there is already fairly stable demand, it is often possible to implement a pull system at an early stage, and this should be an implementation priority. Apart from the advantages mentioned, the pull system often allows simplicity in scheduling since only end-items are scheduled with components simply being pulled as required.

An important requirement for JIT implementation is to establish the links between the Stage One and Stage Two actions. Set-up reductions must translate directly into smaller lot sizes and hence smaller regular schedules. Quality, maintenance and layout improvements must translate quickly into buffer stock removal. Product design rationalisation should allow a regular schedule to be set up for a smaller time bucket. Finally, the compression of lead-time must make the regular schedules increasingly representsative of market rates of demand.

As mentioned, lot-size reduction must go hand in hand with set-up time reduction. But there is another aspect to the lot size concept, and that is the attitude to the holding costs of inventory. The JIT concept is that excess inventory literally hides problems and prevents much improvement. Most importantly, lot size is directly related to lead-time which it is the aim of JIT to reduce, and is directly related to work in process. Therefore, the holding cost of inventory is not simply perhaps the opportunity cost of capital but is much, much higher. If management truly believes that inventory is the worst investment possible, it becomes a management priority to get rid of. This attitude should permeate management thinking on overproduction, safety stocks, storage, and lot sizing. Here JIT implementation is clearly 'in the mind'.

It is dangerous to move towards JIT scheduling unless there is good control over inventories. Many of the prerequisites of MRP – inventory record accuracy, accurate bills of materials, and accurate delivery lead-times – are no less critical with JIT. In more complex environments, a basic MRP explosion will be required. This is not to say than an MRP system is the way to begin. Indeed, the Stage One actions can always be taken whilst simultaneously improving basic inventory control.

Vendor relationships

Vendor relationships are deliberately discussed last because, in contrast to common beliefs about JIT, that is where they usually belong. Working with suppliers is central to JIT realising its full potential, but it is not a prerequisite to begin a JIT programme. As we now know, JIT is a waste reduction and a lead-time reduction strategy, not an inventory reduction technique. As an element of JIT, working with vendors should also be seen as a ongoing programme to reduce supply lead-times and to reduce procurement complexity. The same classification steps described under Stage One are appropriate.

As lot sizes are reduced, it is desirable to have parts delivered in smaller lots just in time for assembly or processing. However, at least initially, the warehouse can fulfil this role. Simultaneously, the buyers must begin the task of working with suppliers to ensure delivery in small lots and with perfect quality. This is a task that may take years, so an early start is desirable. Importantly, there are rewards for both parties. For the supplier, the reward is that he can move towards JIT even faster. He can produce in small regular lots, he can avoid finished goods inventories, and he can enjoy an increasingly secure market.

Under JIT, procurement is a whole new art. Procurement should play the central role between supplier on one hand and design, quality and manufacturing on the other. The buyer becomes a facilitator of communication, spending major amounts of time in suppliers' plants and with his own engineers and operations managers. As with many other areas, JIT implementation requires new attitudes from managers and a new role for the personnel.

Concluding remarks

Just-in-time implementation is an incremental, ongoing process. Unlike many manufacturing systems it is not something that is implemented and forgotten. Rather it is best seen as a strategy which helps people to develop manufacturing skills and capabilities, particularly with respect to problem solving. Moreover, JIT provides a framework in which the ongoing cycle of improvements can be clearly seen.

References

[1] Fukuda, R. 1983. *Managerial Engineering*. Productivity Press, Stamford, Connecticut.
[2] Hayes, R. and Wheelwright, S. 1984. *Restoring our Competitive Edge: Competing through Manufacturing*. John Wiley and Sons, New York.
[3] Schonberger, R. 1982. *Japanese Manufacturing Techniques*. Free Press, Macmillan, New York.
[4] Hall, R. 1983. *Zero Inventories*. Dow Jones/Irwin/APICS, Homewood, Illinois.
[5] Shingo, S. 1981. *Study of Toyota Production System from Industrial Engineering Viewpoint*. Japan Management Association, Tokyo.
[6] Ishikawa, K. 1986. *What is Total Quality Control: The Japanese Way*. Prentice Hall, Englewood Cliffs, New Jersey.
[7] Lu, D. and Ohno, T. 1986. *Kanban: Just-in-time at Toyota*. Productivity Press, Stamford, Connecticut.

STRATEGIES FOR IMPLEMENTING JIT

C. A. Voss and A. Harrison
University of Warwick, UK

In the UK an ever-increasing number of companies are looking to JIT
to improve their manufacturing performance and hence become more
competitive in the marketplace. In a recent survey of UK
manufacturing companies carried out by the University of Warwick, it
was found that 57% of responding companies were implementing or
planning to implement some aspect of JIT. This is not surprising.
Whereas Advanced Manufacturing Technology (AMT) has been
characterised by high investment, high promise but long-term and
uncertain pay-off, JIT has been characterised by low investment, high
short term pay off and high predictability of pay-off. For example, the
first company to be advised on JIT by the Warwick Manufacturing
Round Table at the University of Warwick achieved a 60% reduction in
work-in-progress (WIP), a 98% improvement in quality, and a 20%
improvement in productivity in the first six weeks of operation of its
first pilot cell. Such reports are commonplace.

The data collected in this survey[1] did not always indicate a completely
satisfactory picture of JIT in the UK. First, despite widespread
intentions, there is very little real action taking place. Only 8% of
responding companies stated that they were conducting a major JIT
programme (see Fig. 1). The survey also concluded that many
companies were focusing on easy to implement techniques, and were
neglecting core JIT techniques which would yield far better pay-offs.

Strategic issues in JIT

Strategic issues in manufacturing fall into two categories. The first
addresses those issues that are related to how a company interacts with
customers and competition in the marketplace and how manufacturing
can support a company's competitiveness. The second is concerned
with top management choices and priorities. From our work with many
companies who are considering JIT, five key strategic questions have
been identified:

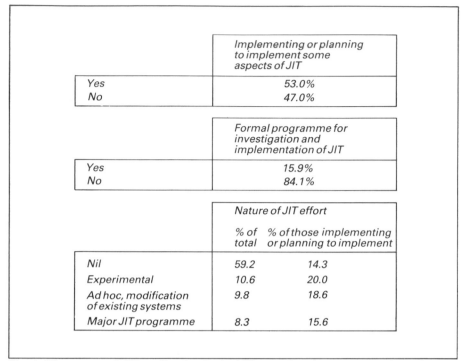

	Implementing or planning to implement some aspects of JIT
Yes	53.0%
No	47.0%

	Formal programme for investigation and implementation of JIT
Yes	15.9%
No	84.1%

Nature of JIT effort		
	% of total	% of those implementing or planning to implement
Nil	59.2	14.3
Experimental	10.6	20.0
Ad hoc, modification of existing systems	9.8	18.6
Major JIT programme	8.3	15.6

Fig. 1   The extent of JIT application in the UK – results of a survey conducted in July 1986

- How will JIT impact on the marketplace?
- How suitable is JIT for our *particular* manufacturing environment?
- Should we invest in JIT, AMT or both?
- How should we implement JIT?
- What fundamental changes do we need to make, to make our company into a JIT company?

This paper will briefly address each of these issues in turn.

JIT and competitiveness

Any company seriously considering its manufacturing strategy must start by considering the marketplace and competition. If your company is competing against the Japanese, it is competing against JIT. The JIT philosophy is now so widespread in Japan, that few exporting companies do not use all or some of the JIT approaches and techniques. JIT is widespread in other companies as well. In West Germany, the University of Passau has helped over 100 companies from a wide range of industries to implement JIT, a number probably far more than the total for the whole of the UK. In the USA, JIT has been widely implemented, particularly following the books of Richard Schonberger. The 'honour roll' in his latest book lists 100 examples of companies

Table 1 JIT and competitive advantage

JIT capability	Competitive advantage derived from JIT capability
WIP reduction	Lower cost manufacture Reduced order to delivery lead time
Increased flexibility	Responsive to customer demands: volume, short lead time, product change
Raw materials reduction	Lower cost manufacture
Increased quality	Higher quality products Lower cost manufacture
Increased productivity	Lower cost manufacture
Reduced space requirements	Lower cost manufacture
Lower overheads	Lower cost manufacture

who have achieved major and quantified benefits from JIT and who are translating these into the marketplace[2].

The benefits from JIT can be considered in terms of their market impact. For example, reduction in manufacturing lead time can lead to being able to compete on responsiveness in the marketplace. Table 1 lists the benefits (in rank order as reported in the Warwick University survey[1], and shows how these can lead to different forms of competitive advantage. The links can also be looked at from the reverse perspective. If a company has chosen its competitive strategy, it can match the strategy to the particular capabilities of JIT. Examples of how this can be done are shown in Table 2.

Matching JIT to the manufacturing environment

The traditional view is that JIT is most suitable for a repetitive manufacturing environment. There is no clear definition of what comprises repetitive manufacturing. Companies usually define it as being manufacture of products that are regularly ordered. One UK company defines this as being any product that is ordered at least once per month by a customer, another as at least once a week.

Table 2 Company strategies and JIT

Competitive strategy	JIT capability supporting strategy
Rapid response to customer needs	Flexibility WIP reduction
Compete on quality	Increased quality
Compete on price	WIP reduction Raw material reduction Increased productivity Reduced space requirements Lower overheads
Rapid product change	Flexibility

Using this definition, repetitive manufacture will embrace most flow and line manufacture, a surprisingly large proportion of batch manufacture but only a limited proportion of jobbing and projects. Methodologies are being developed to determine for a batch manufacturing environment, whether all or part of a company's products are repetitive and hence suitable for JIT[3]. The four main forms of repetitive manufacture are as follows:

- *Family* – different families of parts, each with its own small line.
- *Mixed* – mixed models made on a single line.
- *Dedicated* – single model, single line.
- *Flow* – flow process, chemical, etc.

There are a number of industries where there is already a high degree of flow in manufacturing, such as food and drink, where companies are not sure whether JIT has anything to offer. These flow-based manufacturing environments can be considered as having a large proportion of the elements of JIT in place, however, they can usually benefit by the adoption of selected JIT techniques. In particular, JIT purchasing, and set-up time reduction can lead to major benefits in process industries.

At the other end of the spectrum, the nature of jobbing and projects makes them unsuitable for the flow elements that are the core of JIT. However, as with flow manufacturing, selected JIT techniques can be suitable for these environments. These include set-up time reduction, total quality control and workforce flexibility.

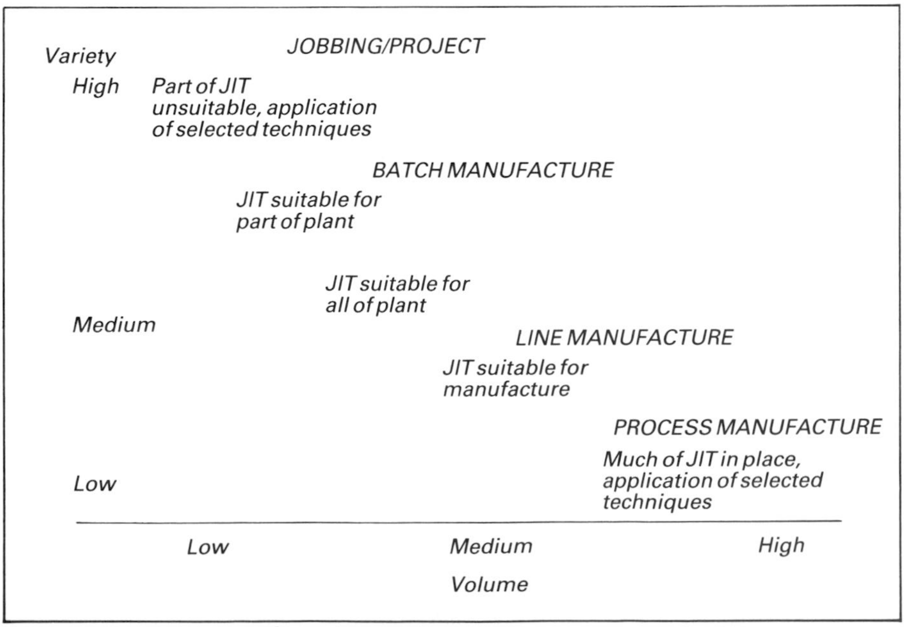

Fig. 2 JIT and the manufacturing environment

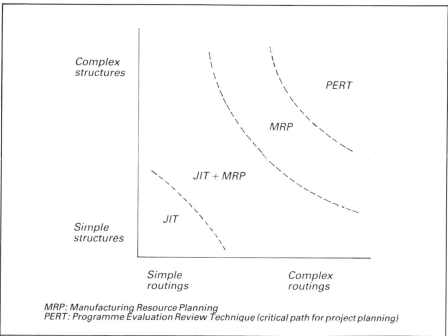

Fig. 3 Suitability of JIT for controlling production

The suitability of JIT for particular process environments is summarised in Fig. 2. The companies in the centre of the diagram, almost certainly have part or all of their manufacturing suitable for JIT, those at the top right and the bottom left will be suitable for selected applications.

A second means of determining whether JIT is appropriate is to examine the inherent complexity of product structure and process routing. The more complex these are, the less suitable JIT is for planning and controlling production. This is illustrated in Fig. 3. However, this is not a static relationship. One of the key elements of JIT is the reduction of complexity (hence the 'lessons in simplicity' described by Schonberger[4]. Companies who are in an environment unsuitable for JIT can make it suitable by a programme of reduction of product and process complexity.

JIT and advanced manufacturing technology

It has been argued by a number of people, for example Ingersoll Engineers, that the route to automation is 'simplify, automate, integrate'. An increasing number of organisations are using JIT as the 'simplify' element in their automation strategies. There are a number of strong arguments that companies considering automation and CIM should pursue JIT in advance of or in parallel to automation. First, some of the methodologies used for planning CIM are used in JIT, in particular group technology. Second, if a major JIT programme is conducted prior to automation, the simplification inherent in JIT may:

- Reduce the amount of investment in automation required.
- Change the nature of the automation required. For example one principal of JIT is to choose many small machines rather than one large one.
- Release capital through WIP reduction to pay for some or all of the automation.
- Reduce the need for computer integration, a JIT plant is physically integrated.

This leads to the conclusion that we should change the Ingersoll dictum to 'JIT, automate, integrate'!

JIT implementation

Successful implementation of JIT requires a number of important decisions to be made. These include what do we start with, where do we start and who should do it? There are probably more than 40 different JIT techniques and approaches. It is difficult to know where to start. We have found it useful to bring together the main techniques of JIT into two groups. In each group, the techniques support each other. The first group, 'JIT 1', is composed of the areas that are necessary for full JIT to work. They focus on four main elements of JIT that can be achieved in the short term: simplicity, flow, quality, and fast set-up. The techniques of JIT 1 are outlined in Fig. 1 of the previous paper in this section by John Bicheno. Without the successful application of these, further progress is difficult. In particular, quality of parts, product and process can be seen as a prerequisite for JIT. A focus of JIT 1 is the process. No company can successfully be a JIT company without full understanding of the process in manufacturing.

Once these techniques and approaches have been implemented and assimilated the more sophisticated and difficult to manage techniques of JIT can be adopted. This group 'JIT 2', includes some of the more well-known ones such as Kanban, and JIT purchasing. (The main elements of JIT 2 are shown in Fig. 2 of John Bicheno's paper.) Many JIT practitioners argue that JIT techniques, in particular JIT purchasing, should not be implemented until the company can handle JIT 1.

Where to start has frequently been answered by choosing a small cell or unit, and once this has been made to work, rapidly transferring the skills and experience round the organisation. This approach was used in Japan, when JIT was being introduced[5], and is being used by many firms in the west such as IBM[6]. IBM argues that the first site should be one where success can be expected, one categorised by a stable manufacturing environment. However, this should not be seen as an ad hoc approach. The initial JIT cell should be the first stage in a long-term plan for implementation.

In organising for implementation, the following approach is frequently used:

- *Steering committee* – This is a cross-functional group of personnel who meet regularly to set objectives for and to direct and monitor the implementation programme. It includes representatives from design, purchasing quality manufacturing, marketing and personnel. It could well include shop-floor representation.

- *Project manager* – It is beneficial to appoint a full-time project manager to act as 'project champion' of the implementation programme. His or her duties include publicising the programme, training, facilitating meetings and 'fixing' money and resources.

- *Project team* – A common form of managing JIT implementation is to set up implementation teams. These teams may well be shop-floor based, rather than a head office firefighting team. The project manager helps to set up these teams and to establish terms of reference. Teams should be cross-functional, but related to the particular techniques being implemented. Teams should be given total backing to implement change without delay, this is the spirit of JIT.

  Teams are trained by various means such as visiting successful users, using consultants, in-company course, courses run by institutions such as the University of Warwick and the use of videos. Implementation needs to be accompanied by a number of far-reaching changes in the organisation. These are discussed in the next section.

Becoming a JIT company

There are many things embodied in the traditional way of manufacturing which conflict with the requirements of JIT. It is not surprising that the main resistance to JIT often comes from middle level management and technical specialists who are constrained by today's methods of management. There are four areas where a company must change:

- *Measures of performance* – JIT demands a new look at measures of performance. Traditionally companies have tried to maximise output and machine utilisation. In a JIT environment, the short objective is to meet planned output (and not exceed it), and to have use of undercapacity scheduling, which may result in having planned spare capacity. In a JIT company it is important that in addition to measures of productivity and cost, management focuses on the following measures as being vital manufacturing performance measures: WIP level, quality, manufacturing lead time, distance travelled, space utilisation. Using traditional measures alone, will militate against successful JIT Implementation. The focus has moved away from cost minimisation to waste elimination and the maximisation of value added.

- *Flexibility in the organisation* – To make JIT work requires considerable flexibility in the organisation. This is not just at the shop-floor level but throughout line and staff functions. This is a long and difficult road, so companies adopting JIT must start considering how they will make their organisations more flexible from the very start.

- *A focus on the doer* – In implementing JIT more responsibility is pushed down the line to the doers. They are closer to how things work and the associated problems. For example, many companies have built small self-contained cells or modules. Within these, the team have wide-ranging responsibility for organising their own work, quality control, maintenance and repair of their own machines and movement of material to and from their cell. Tasks which have been traditionally performed by staff and/or indirect functions are now performed by the doers. Delegation from specialists (IEs, quality inspectors, etc.) should be enforced wherever possible.

- *Continuous improvement* – The traditional focus of manufacturing management has been problem solving. We are good at this; however, once a problem is solved we tend to leave well alone. In a JIT company, the focus changes to continuous improvement. When a problem is solved we must seek to improve still further, we must seek for further ways to eliminate waste. JIT is a journey of continuing improvement.

These are four fundamental changes in the ways companies operate, and have been achieved not only in Japan, but by a number of UK companies. They require a major commitment by the company at all levels. Once achieved, these new ways of managing allow for new orders of magnitude of performance.

References

[1] Voss, C.A. and Robinson, S.J. 1987. The application of JIT in the UK. *Int. J. Operations and Production Management*, forthcoming.
[2] Schonberger, R. 1986. *World Class Manufacturing*. The Free Press Division of MacMillan, New York.
[3] Berry, W.L., Tallon, W.J. and Wolfs, D. 1986. Manufacturing Planning and Control Systems: A Key Competitive Element in the Formulation of Manufacturing Strategy. Unpublished working paper, School of Business, University of Iowa.
[4] Schonbergr, R. 1985. *Japanese Manufacturing Techniques: Nine Hidden Lessons in Simplicity*. The Free Press Division of MacMillan, New York.
[5] Nakayama, Y. 1986. JIT in the factory. In, *Proc. CBI JIT Conf.*, Birmingham.
[6] Stride, P. 1987. JIT at IBM. Presentation made at the *UK Operations Management Association Conf.*, Nottingham.

5

JIT PURCHASING

JIT extends beyond the boundary of the firm to both suppliers and customers. This section looks at the new approaches to purchasing that have been developed originally in Japan, and subsequently used by Western companies implementing JIT purchasing and supply. Topics examined include supplier resource management, supplier selection, delivery requirements, performance monitoring and feedback, sourcing policies, receiving at the point of use and quality issues. In addition there are comments on the changing role of the buyer in a JIT purchasing environment.

COMPARING JAPANESE AND TRADITIONAL PURCHASING

S.M. Lee
University of Nebraska Lincoln, USA
and
A. Ansari
Seattle University, USA

Just-in-time (JIT) purchasing has received an increasing amount of attention in the operations management literature. Today, a growing number of US firms have switched to the Japanese JIT purchasing concept in an effort to improve their product quality and productivity. This paper discusses the major activities of JIT purchasing and provides a comparative analysis of differences between the JIT purchasing and traditional US purchasing systems. Furthermore, the reasons behind these major differences and their future implications for US firms are discussed.

In the early 1950s, a unique production system emerged in Japanese manufacturing companies which contributed substantially to Japan's high product quality and productivity. During the past two decades, Japan's annual productivity increase rate in manufacturing was 9.3% as compared with a 2.7% increase in the USA. This distinctive Japanese system is widely known as Just-in-time (JIT) production. An important aspect of JIT that has had a great influence on product quality and productivity is JIT purchasing.

JIT purchasing is an important element of this unique Japanese production planning and inventory control technique. JIT purchasing is effective for the following:

- Controlling the inventory system.
- Reducing buffer inventories.
- Reducing space needed.
- Reducing material handling.
- Reducing wasted materials.

During the past several years, practitioners and academics who were concerned with product quality and productivity in the USA have focused increasing attention on the potential benefits of the JIT concept. Since the early 1980s, many American and Japanese executives and scholars have participated in regional, national, and international conferences to exchange views on business, economic, and technological advances. One of the most popular topics at these conferences has been the implementation of Japanese manufacturing techniques.

Today, a growing number of US companies has switched to JIT purchasing from traditional purchasing practices. A few of these companies have implemented their own versions of JIT under different names, such as ZIPS (zero inventory product system), MAN (material-as-needed), and nick-of-time. By any of these names, JIT treats purchasing the same way[1]: materials are purchased in small quantities with frequent deliveries from a fewer number of close suppliers, just in time for use. It appears certain that during the next few years there will be an accelerated level of interest among many US companies in the implementation of the JIT purchasing concept.

Through interpreting data collected during a recent study this paper seeks to:

- Identify the major activities involved in JIT purchasing.
- Compare these activities with traditional US purchasing.
- Discover the reasons behind the differences.
- Discuss future implications for US firms.

Data collection involved three approaches: questionnaire responses, interviews and collected documents. The first utilised a questionnaire to collect the data in seven categories: company descriptions, transportation/traffic details, vendor relations, quality inspection, benefits of JIT purchasing, implementation problems, and prerequisite variables required for successful implementation.

The second approach consisted of interview with the purchasing manager, production manager, quality control manager, engineering and design people, and the transportation/traffic manager in four US companies: (1) General Motors Corporation (Buick Division); (2) Hewlett-Packard Company (Greeley Division); (3) Nissan Motor Manufacturing Corporation USA (Smyrna plant); and (4) Kawasaki Motors Corporation USA (Lincoln plant). This provided empirical on-site data.

The third method of data collection utilised documents collected from companies and professional associations which are proponents of the JIT effort in the USA, e.g. Automotive Industry Action Group. The documents collected were used to support management responses from the interviews as well as the responses to the questionnaire.

JIT purchasing versus traditional US purchasing

Purchasing activities include all of the functions involved in the procurement of material, from the time of need to receipt and use of the material. These activities vary considerably among organisations depending on the size of the company. Generally, there are several major activities over which purchasing should have full responsibility, regardless of the organisation's size.

The major activities of purchasing practice are as follows:

- Determining the purchase lot size.
- Selecting suppliers.
- Evaluating suppliers.
- Receiving inspection.
- Negotiating and bidding process.
- Determining mode of transportation.
- Determining product specification.
- Paperwork.
- Packaging.

Table 1 Comparative analysis of purchasing practice: Traditional US and Japanese JIT

Purchasing activity	JIT purchasing	Traditional purchasing
Purchase lot size	Purchase in small lots with frequent deliveries	Purchase in large batch size with less frequent deliveries
Selecting supplier	Single source of supply for a given part in nearby geographical area with a long-term contract	Rely on multiple sources of supply for a given part and short-term contracts
Evaluating supplier	Emphasis is placed on product quality, delivery performance, and price, but no percentage of reject from supplier is acceptable	Emphasis is placed on product quality, delivery performance and price but about 2% reject from supplier is acceptable
Receiving inspection	Counting and receiving inspection of incoming parts is reduced and eventually eliminated	Buyer is responsible for receiving, counting, and inspecting all incoming parts
Negotiating and bidding process	Primary objective is to achieve product quality through a long-term contract and fair price	Primary objective is to get the lowest possible price
Determining mode of transportation	Concern for both inbound and outbound freight, and on-time delivery. Delivery schedule left to the buyer	Concern for outbound freight and lower outbound costs. Delivery schedule left to the supplier
Product specification	'Loose' specifications. The buyer relies more on performance specifications than on product design and the supplier is encouraged to be more innovative	'Rigid' specifications. The buyer relies more on design specifications than on product performance and suppliers have less freedom in design specifications
Paperwork	Less formal paperwork. Delivery time and quantity level can be changed by telephone calls	Requires great deal of time and formal paperwork. Changes in delivery date and quantity require purchase orders
Packaging	Small standard containers used to hold exact quantity and to specify the precise specification	Regular packaging for every part type and part number with no clear specifications on product content

All of these activities in traditional US purchasing are approached differently from JIT purchasing practices. Table 1 is a summary comparing activities under both systems. Activities are elaborated below.

Purchase lot size

Traditional US purchasing practices generally rely heavily on a just-in-case system. Large batches are purchased based on a delivery schedule specified in the purchase order. This practice allows the company to continue to operate even if there are serious disruptions of supplies[2]. On the other hand, under Japanese JIT purchasing practice the emphasis is placed on the purchase of minimum lot sizes, preferably piece for piece. Piece-for-piece delivery allows a tighter inventory control which eliminates large stocks of parts between process stages[3,4].

One of the most important justifications for the traditional purchasing practice of buying in large quantities with less frequent deliveries rather than in small quantities with more frequent deliveries is the realisation of lower shipping and handling costs. Schonberger[4], however, argues that most US companies purchase in large lot sizes because they consider the shipping and handling costs as given items. Under the JIT purchasing concept, obtaining small lot sizes is considered a challenge in spite of obstacles such as shipping costs.

There are two approaches[5] which have been used by many US companies, such as Kawasaki (Lincoln Plant), Hewlett-Packard (Greeley, Colorado Division), and General Motors (Buick Division), to reduce JIT shipping costs associated with the frequency of delivery. The first approach is to develop a freight consolidation programme with various suppliers. One of the important benefits of this programme is to aid the suppliers to share delivery trucks. In this arrangement, it is not uncommon to see trucks carrying small quantities of parts from three or four suppliers. Each truck is assigned a time when it can enter the plant and drive directly to the assembly line. The second approach is to select local suppliers, wherever possible. When there is no local source, potential suppliers are encouraged to move their operations close to their plants through offers of long-term contracts.

Selecting suppliers

The most important activity of purchasing is supplier selection. A key feature of JIT purchasing practices, which differs from traditional US purchasing practices, is the idea of dealing with a small number of nearby suppliers – ideally single sources of supply. Under JIT purchasing, the buyer is encouraged to buy a given part from a single supplier in a nearby geographical area and to establish good long-term relationships with the supplier. Edward Hennessy[6], CEO at Allied Corporation, argues that:

"Purchasing must cultivate sound relationships with its suppliers so inventories may be reduced to minimum practical levels and quality of supply may be such that rejection of material is esentially eliminated."

Single sourcing is contrary to the typical US purchasing practice which generally relies on multiple sources of supply for a given part. Deere and Company indicated that a few years ago it had many suppliers for a given part and had to split the order among them. For example, they purchased forgings from 50 different companies[7].

Strong arguments arise between those who advocate single sourcing and those who support multiple sourcing. The proponents of multiple sourcing argue that multiple sources of supply give broader advantages to the buyer, such as:

- It provides a broader technical base to the buyer.
- It protects the buyer in times of shortages against failure at the supplier's plant.
- It encourages competition among suppliers in securing the best possible price and products[8].

On the other hand, those US companies that have implemented the JIT purchasing concept realised that having a smaller number of sources of supply results in certain benefits:

- Consistent quality – involving suppliers in the early stages of product design can consistently provide high quality products.
- Saving of resources – minimum investment and resources, such as buyer's time, travel, and engineering that are required when there is a limited number of suppliers.
- Lower costs – the overall volume of items purchased is higher, which eventually leads to lower costs.
- Special attention – the suppliers are more inclined to pay special attention to the buyer's needs since the buyer represents a large account.
- Saving on tooling – total dollars spent to provide tooling to the supplier is minimal since the buyer concentrates on one source of supply.
- The most important benefit obtained from single sourcing is the establishment of long-term relationships with suppliers, which encourages loyalty and reduces the risk of an interrupted supply of parts to the buyer plant.

Evaluating suppliers

The evaluation of sources of supply is the most important continuing process of purchasing. Evaluation varies according to the nature, complexity, competition, and dollar value of items purchased. Under traditional US purchasing practices, evaluation is based on three important criteria[9]: product quality, delivery performance and price. Determining which of these factors is the most critical depends on how their relative importance is perceived. A study conducted among 273 purchasing managers[10] listed 23 factors that were considered in evaluating a potential supplier. This study revealed that product quality and delivery performance were extremely important factors and price was considered merely as an important factor.

One effective method available in evaluating a supplier's quality of performance is to make monthly or quarterly tabulations of the percentage of rejected materials. Another method for rating a supplier's quality performance involves a regularly scheduled review of quality performance among the buyer, suppliers, and the engineering people[8].

Delivery performance includes responses to enquiries, special services rendered, and other intangibles. It is common for US companies to develop a vendor delivery rating scheme that consists of categories, such as top rating, good, fair, and unsatisfactory. The supplier's delivery performance is then tabulated and rated monthly.

Although traditional US purchasing practices and Japanese JIT purchasing practices both emphasise the importance of product quality and delivery performance, emphasis placed on these factors varies between the two approaches. For example, in contrast to the traditional US purchasing practice, where many companies accept about 2% rejects from suppliers, Japanese JIT purchasing permits no such percentage of rejects because the supplier has the responsibility to deliver just the right number of items. When Hewlett-Packard recently asked for a large sample of a component part, it subjected its potential suppliers to intense reliability testing. The best Japanese supplier had 0.003% unreliability whereas the best American supplier achieved 1.8% unreliability[11].

Receiving inspection

Under traditional US purchasing practice, the receiving department is responsible for receipt, identification, piece-by-piece counts and inspection of all inbound freight for quality in accordance with their product design specifications. The responsibility of inspecting the incoming products is almost invariably placed on the buyer. In the Japanese system, receiving inspection is avoided except for new parts and new suppliers[12].

Under JIT purchasing, in most Japanese plants, it is common for suppliers to drive their trucks straight to the assembly line (except for new parts and new suppliers)[13]. This practice is primarily achieved by extending the inspection function of quality back to the supplier's plant and making sure that quality is built-in before the part leaves the supplier's plant.

Many US companies are in the process of moving the responsibility for quality of all incoming parts back to the suppliers. This concept has already been adopted by Nissan (Smyrna plant), GM (Buick Division), Kawasaki (Lincoln plant) and Hewlett-Packard (HP) for some of their potential JIT suppliers. For example, ComMents (CMT), one of HP's suppliers, moves its delivery cart directly to the production area without going through inspection, avoiding excessive paperwork, etc.[6].

Negotiating and bidding process

Another difference between traditional US and JIT purchasing practices is the negotiating and bidding process. Since the typical US buyer preference is to deal with multiple sources of supply, the traditional bidding process implies that the lowest bid customarily will get the contract. In fact, the main objective in bidding from various sources is to obtain the lowest possible price. The primary reason for this is that most US buyers provide very exact and rigid product specifications for prospective suppliers, and the buying decision is therefore usually based on lowest cost. An important aspect of negotiating contracts is that the supplier offers very short-term contracts and they may be terminated for reasons of competitive price.

In contrast, the objective of bidding under JIT purchasing practices is not just to negotiate for the lowest bid possible, but also to establish a close relationship with the suppliers for the following reasons:

- The concept of JIT purchasing emphasises a single source of supply. Therefore, the buyer and supplier will agree upon a 'fair' price to both parties.
- The bidding specifications are not as rigid, and suppliers are encouraged to be innovative in meeting the specified needs.
- Emphasis on product quality is the primary factor in the bidding and negotiation process.
- An increase in the number of longer term contracts with the possibility of annual negotiation with the supplier, leading to better quality and possible cost reduction.

Determining mode of transportation

Transportation is an important issue in most manufacturing firms. Due to complexities involved in transportation, responsibility is often placed in a separate department designated as the traffic department. The responsibilities of the traffic department include outgoing freight and internal plant transportation.

In traditional US manufacturing buying, the handling of inbound freight differs from JIT buying. According to US practice, the primary responsibilities for scheduling and delivery are generally left to the supplier and the transportation company, regardless of whether the purchase contract states FOB destination or FOB shipping point. The emphasis for US manufacturing traffic managers is therefore on outbound freight. The reasons why many traffic departments do not concentrate on inbound freight is that most industrial traffic departments are measured on how successful they are in lowering outbound transportation costs.

Under JIT buying, the traffic manager is responsible for both inbound and outbound freight. Schonberger[4] states that:

"JIT buying can hardly be successful if inbound freight scheduling is left to the transportation system, whose primary concern is with optimal utilisation of drivers, storage space, and trailer or rail-car cubes."

Under JIT, traffic managers are more concerned with on-time delivery than with trying to lower outbound costs. They also try to prevent production disruptions in the buyer's plant due to late arrival of goods.

Product specification

According to traditional US purchasing practice, the engineering department specifies and develops the tolerance for almost every conceivable design feature of the end product[14]. At the same time, purchasing reviews purchase requirements to make sure that any necessary product specifications (specs) are defined. Suppliers know exactly what the buyer wants[15]. Several types of specs commonly used in the US are blueprints, performance specs, material specs, etc. The buyers rely more on design specs (which describe and identify the composition of materials to be used, their size, shape, and method of manufacture), than on performance specs. Although the design engineers are responsible for developing these specs, they rarely interact with suppliers. Procurement problems are left to the purchasing department. This reduces the feedback that design engineers will receive from suppliers in the area of design or quality.

Under the JIT purchasing practice, the buyer seeks more technical advice and assistance from suppliers in order to design better parts, achieve lower prices and improve product quality and productivity. The buyer relies more on supplier performance specs and less on narrowly defined design specs. This allows the supplier to use more discretion in making recommendations and innovations when discussing with the buyer any problems with respect to design and quality. Generally, relying on performance specs places greater responsibility on the supplier for satisfactory products. Additionally, work delays are avoided in the supplier's plant because there is more freedom in dealing with product design[4].

Paperwork

Under traditional US purchasing practice, purchase orders are issued for nearly all requirements such as purchase requisitions, packing lists, shipping documents, invoices, etc. These activities and their supporting documents require a massive amount of time and formal paperwork. Janson[16] explained that in traditional purchasing the paperwork is so extensive that purchasing people spend most of their time pushing paper. Wight suggested that frequent changes in order quantity and delivery times forced purchasing people towards a reactionary 'fire-fighting' mode; more than 50% of their time was spent on paperwork and expediting. This gave purchasing people less opportunity to 'problem-solve' with suppliers and to work closely together to improve cost efficiency, product design specs, and productivity[17].

JIT purchasing requires much less formal paperwork since: deliveries are made several times a day; long-term contracts are used; and a simple phone call can easily change the delivery timing and quantity level. Also, paperwork is reduced under the JIT system by use of Kanban cards which trigger the deliveries. Therefore, purchasing has more time available to spend with the suppliers to improve the product design specs, quality and productivity. For example, Berry[18] explained that Newman Foundries uses its own trucks to deliver aluminium castings to Chevrolet transmission once or twice a day. Newman "may be one of the first suppliers to deliver to GM without paper documentation."

Packaging

A factor which is often overlooked in traditional US purchasing practices, but with which the buyer should be concerned, is packaging specification and handling. Better packaging and precise specifications of product content not only reduce manpower requirements, but also affect the distributor, retailer, marketing department, and transportation department[15]. Ammer[19] suggested that purchasing should be concerned not only with the flow of materials into the plant and finished products out of the plant, but also with product specification and handling, which might be quite complex and costly.

Aljian[9] indicated that:

"Packaging improvements may consist of such a small thing as specifying smaller containers to permit one-man handling or to prevent losses due to opened, partially emptied containers which permit loss, deterioration, or contamination."

Under JIT purchasing the idea is to use small standard containers for every part type and part number. Since the containers hold a precise quantity, the following advantages are realised:

- No overage/underage allowed.

- Precise specification of parts on the containers prevents the buyer from making mistakes.
- It allows easy and accurate count of parts.

MacCallum[20] explained that Honda's suppliers in Japan often use plastic bins to deliver the parts directly to the assembly line. John Stalk, Materials Operations Manager at Xerox[21], explained that under its new programme, Xerox aims to eliminate defective parts through several factors. One of the important factors is better packaging and handling which should arise from a closer relationship with suppliers.

Implications for US firms

The contrast between Japanese purchasing practices and US traditional purchasing practices provides a good insight into the relative importance of JIT for US firms. Japanese automobiles, television sets, cameras, robotics, ceramics, optics, and other products have captured a sizeable portion of the US market. Many authors believe that Japan's success in achieving a high rate of product quality and productivity is largely a result of its manufacturing techniques, especially the JIT system

Since the initial implementation of JIT purchasing in late 1980 by Kawasaki Motors (Lincoln plant), at least 50 companies in the US such as GM, Ford, Nissan, Hewlett-Packard, Xerox, Goodyear, GE, and other large US corporations, have adopted and implemented the concept in their plants. The implementation of the JIT purchasing concept has significantly altered not only the way these companies plan and control their production facilities and purchasing systems but has also led to improvements in their product quality and productivity[5].

The major activities of JIT purchasing which substantially vary from traditional US purchasing are:

- Purchase lot size – materials are purchased in very small quantities with frequent deliveries.
- Selecting supplier – single sourcing and long-term contract.
- Evaluating supplier – emphasis on quality with no percentage of reject acceptable.
- Receiving inspection – the need for incoming inspection can be reduced.
- Negotiating and bidding process – high quality and fair price.
- Transportation – delivery schedule (inbound and outbound freight) left up to buyer.
- Product specs – use of less rigid material specs and encouragement of more innovation on the supplier's part.
- Paperwork – eliminate unnecessary paperwork.
- Packaging – utilise small standard containers to hold exact quantity.

References

[1] Waters, C.R. 1984. Why everybody's talking about 'Just-in-time'. *INC*, 6(3): 77–90.

[2] Nellermann, D.O. and Smith, L.F. 1982. 'Just-in-time' vs just-in-case production inventory systems: Concepts borrowed back from Japan. *Production and Inventory Management*. 23(2): 12–20.

[3] McElroy, J. 1982. Making Just-in-time production pay off. *Automotive Industries*, 162(2): 77–80.

[4] Schonberger, R.J. 1982. *Japanese Manufacturing Techniques: Nine Hidden Lessons in Simplicity*. The Free Press, New York.

[5] Ansari, A. 1984. An empirical examination of the implementation of Japanese Just-in-time purchasing practices and its impact on product quality and productivity in US firms. Unpublished dissertation. University of Nebraska Lincoln.

[6] Secrets of Japanese success. *Management Today*, January 1981: 64–68.

[7] New thoughts on purchasing. *Implement and Tractor*. October 1982: ME-7-ME-22.

[8] Lee, L. Jr. and Dobler, D.W. 1977. *Purchasing and Materials Management*, 3rd edition. McGraw-Hill, New York.

[9] Aljian, G.W. 1973. *Purchasing Handbook*, 3rd-edition. McGraw-Hill, New York.

[10] Dickson, G.W. 1966. An analysis of vendor selection systems and decisions. *Journal of Purchasing*, 2(1): 5–17.

[11] Tregoe, B.B. 1983. Productivity in America: Where it went and how to get it back. *Management Review*, 72(2): 23–45.

[12] Schonberger, R.J. and Ansari, A. 1984. Just-in-time purchasing can improve quality. *Journal of Purchasing and Materials Management*, 20(1): 2–7.

[13] Hartley, J.R. 1981. The world's greatest production line: The Japanese view. *Automotive Industries*, 161(12): 53–54.

[14] Schonberger, R.J. and Gilbert, P.J. 1983. Just-in-time purchasing: A challenge for US industry. *California Management Review*, 26(1): 54–68.

[15] Tersine, R.J. 1980. *Production/Operations Management: Concepts, Structure, and Analysis*. Elsevier North Holland, New York.

[16] Janson, R.L. 1980. *Purchasing Agent's Desk Book*. Prentice-Hall, Englewood Cliffs, New Jersey.

[17] Morgan, J.P. 1980. MRP breaks with past patterns of failure. *Purchasing*, 89(2): 48–54.

[18] Berry, B.H. 1982. Detroit automakers slim down inventory to beef up profits. *Iron Age*, 225(24): 61–65.

[19] Ammer, D.S. 1980. *Materials Management and Purchasing*. Richard D. Irwin, Homewood, Illinois.

[20] MacCallum, F. 1982. *Japanese Production Methods and Strategies*. GM of Canada, St. Catherines.

[21] Quality rests on an active supplier. *Purchasing*, 92(2): 97–100.

JIT PURCHASING IN THE UK

V. Winn
Hewlett-Packard, UK

The Bristol Division is Hewlett-Packard's UK manufacturing site for computer peripherals – specifically mass storage devices for the European market. The Division started manufacturing in a temporary facility and shipped the first product in April 1983. Many US divisions were operating in a Just-in-time environment at that time and were experiencing major benefits in areas of inventory, quality, space utilisation, cycle times, etc.

With the Bristol operation being a 'greenfield' site, the company had an ideal opportunity to implement JIT from day one. In September 1984 the operation moved from the temporary 50,000ft^2 building to a new, purpose-built, 125,000ft^2 building in Stoke Gifford, Bristol. This building was the first of a possible nine on a 165 acre site. The second 125,000ft^2 building opened in August 1986, and currently the total headcount, including HP's UK research laboratories and a computer support group is 470 (315 in the manufacturing division).

There are four major product lines producing 14 different products, varying in volume from five units per day to 70 units per day. To support production, the company uses around 1600 active part numbers from about 170 suppliers.

At start-up, creating a JIT environment within the factory did not present too many problems; expertise and experience were available within HP in the areas of line layout, material flow, inventory control, lead/cycle-time reduction, systems implementation, training, etc. The biggest problem was to source suppliers that could consistently meet HP's expectations of quality, on-time delivery and service to support JIT needs. Implementing JIT within the factory is easy compared to JIT outside and to the factory!

Supplier resource management
Selection

Selecting the right supplier and managing that supplier is key to, and the most important part of, the JIT philosophy. At Hewlett-Packard, substantial engineering and commercial effort has been invested into supplier resource management. The first step is to select the right supplier; not simply the lowest price, but the supplier who provides the lowest total cost of ownership in terms of price, cost of quality and ability to deliver on time, every time. With 'stockless' production, poor quality and late delivery is a recipe for disaster!

Among other things in selecting a good supplier, HP looks for internal process control and measurement, competitive pricing, good internal control of procurement, planning and despatch, short and flexible lead-times and, if possible, close proximity to the manufacturing site (see Fig. 1).

Communication of expectations

After careful selection, it is essential to communicate to the supplier the expectations of quality, price, delivery and flexibility. At HP we run a series of seminars for our suppliers to communicate these expectations. The seminar lasts one day and as well as stressing the company's expectations of a supplier, HP explains its manufacturing philosophies and techniques that generate these expectations. The theme of the day is to promote and develop a partnership, as opposed to an adversarial relationship between HP and the supplier. HP is looking for a long-term relationship where both parties benefit.

What does HP expect from a supplier?

- *Lowest total cost of ownership (not just lowest quote)*
- *Parts that work 100% in our process*
- *On-time delivery (every time)*
- *Low consistent lead times*
- *Weekly/daily shipments*
- *Good internal process control*
- *Cost reductions passed on to HP*
- *Good communication*
- *Control of internal planning activities (purchasing, production planning, despatch etc.)*
- *Financially sound*
- *HP business between 10%-25% of turnover*

Fig. 1 Selecting the right supplier

Developing the right relationship

Having selected the supplier and communicated expectations, HP has an on-going programme of monitoring and feedback of supplier performance. Systems have been developed to produce quality and on-time delivery information in the form of graphs and reports.

The quality expectations are that 100% of parts work in the application. At first sight this may seem a fairly loose definition of quality, but as HP does not, in general, perform incoming inspection, it relies on internal processes to uncover any defects. HP's delivery window for suppliers is up to three days early and zero days late, compared to the due date. The due date is delivery on to the receiving dock. Internal systems capture delivery data, compared with due date, and from this automatically produce supplier-specific delivery performance data. Similarly, quality data on defective parts from HP's own process control is fed into this system and provides supplier-specific quality data.

Performance measurement

Other aspects of HP's suppliers' performance are monitored, such as communication effectiveness, technical support, flexibility, etc. and, of course, price competitiveness. All of these performance characteristics are combined and weighted to allow two things to be done: give regular supplier-specific performance feedback and to produce supplier data by commodity to compare one supplier with another.

Feedback to suppliers and on-going management

Supplier performance feedback is given quarterly, but more often if there is a specific problem to be addressed. Supplier management in HP is the responsibility of the materials department within the manufacturing group. Supplier feedback is a teamwork effort by an engineer and a buyer within this department who both work with the supplier on an on-going basis and have developed relationships with personnel within the supplier's organisation. Effective management of the supplier requires consistent, reliable and defined communication channels. HP's materials organisation is structured to optimise this communication. Good communication is essential, not only to provide performance feedback, but also to ensure an early warning system for potential problems in availability or quality. If aware of potential problems one can often plan around them, develop contingencies or provide engineering or commercial assistance towards solving the problem before it occurs. Materials engineers in HP are given responsibility by commodity and are experts in the supplier's process. Likewise, the buyer will be totally familiar with the commercial aspects of the supplier's business (see Fig. 2).

- *Establish good communication channels*
- *Maintain visibility of potential problem situations*
- *Technical experts in supplier's process*
- *Experts in commercial aspects of supplier's business*
- *Provide technical/commercial assistance when appropriate*
- *Maintain supplier base to a manageable level*
- *Control and limit the number of new parts generated during design*

On-going supplier management requires an organisation structure that provides total ownership for all aspects of that activity

Fig.2 On-going management

Sourcing policy

The sourcing policy adopted in HP is to generally single-source parts but multi-source commodities. For example, there may be five sheet metal assembly suppliers, but a specific part would be produced by only one of those suppliers.

This sourcing policy exists for three reasons. Firstly, it is part of the overall, long-term commitment to the relationship with the supplier. Secondly, it allows HP to focus resources on selecting, developing and monitoring one source only. Thirdly, there are some sound commercial reasons for doing so – leverage can be gained from volume in both price and service, there is a greater likelihood of obtaining consistent quality levels, and any tooling investment is limited to one source.

There is, of coure, some associated risk to this policy; when the supplier has problems, there is no second source to help out. There may also be some competitive advantage to having multiple sources. However, it is felt that the advantages of single sourcing outweigh these potential disadvantages.

It is fairly obvious, as stated earlier, that to manage the supplier to this level consumes considerable resources within the organisation. However, it is absolutely necessary, to exist in a JIT environment, that these resources are made available and the people have adequate training, experience and qualifications to enable effective management of the supplier.

JIT production at HP Bristol

Like any other company, HP does not have limitless resources in terms of people, and therefore has an objective to maintain its supplier base to a manageable level. Likewise, it has an objective to minimise the number of discrete parts that the company purchases. Only by having a small base of parts and suppliers and closely managing those suppliers can HP maintain a supply of good quality parts, Just-in-time to support its production.

- Repetitive, demand-pull production system

- No works orders or kitting

- Kanbans limit in-process inventory

- Stores located at workstation

- Daily rate assembly

- No WIP tracking

- Post-deduction of inventory

Fig. 3 Inventory control at Hewlett-Packard, Bristol

Within HP's production areas, a repetitive demand-pull system is operated as opposed to a batch-orientated works order 'push' system. Inventory is distributed within the production areas and located at the workstation where it will ultimately be used. There is not a centralised parts warehouse, and kitting for production is not performed. Assembly is daily rate based and parts are used on an 'as-required' basis (see Fig. 3).

A 'dock-to-stock' system is operated, i.e. when parts are received they are counted and moved directly to the production storage location. Material flow is much simplified and handling drastically reduced. Subassemblies are also built at daily rates to match the product level assembly rate. Once inventory is pulled from stores into process it remains there until the final product is moved into finished goods inventory. 'Kanbans' limit the amount of work in process at any one time (see Figs. 4 and 5).

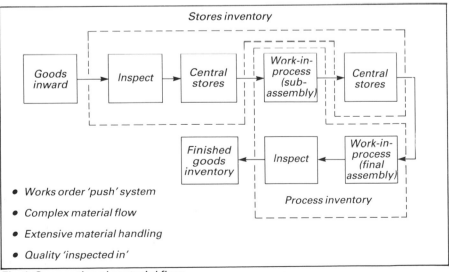

- Works order 'push' system

- Complex material flow

- Extensive material handling

- Quality 'inspected in'

Fig.4 Conventional material flow

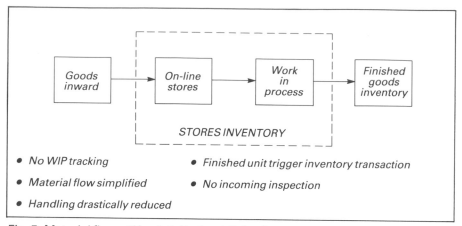

Fig. 5 Material flow at Hewlett-Packard, Bristol

The obvious 'missing link' in the material flow process is incoming inspection. At HP, the approach taken is that quality is the responsibility of the supplier and, as mentioned earlier, the company works very closely with the supplier to ensure good quality output from its process, and relies on HP's own process control to detect defects.

There are exceptions to this: HP will inspect the first two batches of new parts and if the yield is 100% then inspection is removed. If a history of quality problems exists parts will be inspected for a short period, working with the supplier, and inspection will cease after two good batches. For regulatory reasons, some parts are tooled through inspection, but mainly to check standards marking, certificate of conformance, etc.

In the past, mainly during development phases, source inspection has been used as a means of guaranteeing quality levels, but this is very much the exception. For some commodities the supplier is required to provide inspection/test results either with every delivery or on an audit basis.

HP's overriding philosophy is that quality is the responsibility of the supplier and good quality should be the result of a quality process as opposed to 'inspecting-in' the quality.

Shop-floor control

Control of inventory accuracy on the shop-floor demands close attention in a JIT production system, since parts are not locked away and closely guarded by stores personnel. The method used by HP is to cycle count parts regularly; the frequency of counting being dependent upon the usage and value of the part. Cycle counting is performed by a materials handler from the materials department who is located on the production area and who is an integral part of the production team. This individual also performs, on line, all of the inventory transactions to decrement inventory from the computer records for both planned and unplanned activities. There is no distinction between stores location inventory and process inventory as far as computer records of the available inventory is concerned. One transaction, at the end of the production process decrements the complete bill of material (purchased parts level) from the available inventory. Hence cycle counting must include process inventory.

Inventory levels

Inventory level at HP Bristol is maintained around one month's supply, i.e. inventory is turned approximately twelve times every year. This is maintained by careful delivery scheduling; the high value, high usage parts are delivered weekly and in some cases daily. Active parts are divided into 'A, B and C' classification where the 'A' parts account for around 10% of the quantity, and 90% of the value. Clearly, close control of the 'A' parts effectively controls overall inventory value. 'B' and 'C' class parts have some level of safety stock to allow the management of the 'A' class parts extremely tightly. The average value of total inventory (stores, in process, finished goods) held at Bristol Division is about $4 million (one month's supply). Five years ago, a typical Hewlett-Packard division in a conventional production environment, would have held around four month's supply of inventory. The savings in company assets are staggering! The other saving of course is in land and building space to store inventory. The author estimates that with four times the present inventory levels, HP would have needed the second building at the Bristol site (at a cost of $10 million) 12 to 18 months before it actually did (see Fig. 6).

With inventory levels reducing, the need for better communication between the purchasing department and production is increased. It is extremely important that there is visibility within purchasing of parts shortages or potential shortages in production to enable rapid reaction and solution to the problem. At HP Bristol, buyers and production supervisors meet and review parts availability daily. Of course, there are back-up pre-shortage reports to support and enhance the communication.

- *One month's supply (stores, process, finished goods)*

- *Daily or weekly deliveries of high-value usage items*

- *10% of parts account for 90% value*

- *Other 90% of parts have some level of safety stock/flexibility stock*

SAVINGS:

- *$12-$16 million investment in inventory*

- *Delay $10 million investment in second building by 12 months*

- *Overall saving in manufacturing cost*

Fig. 6 Inventory levels

- Inventory control on a defined range of parts
- Supplier management
- Parts availablity management
- Contract negotiation
- Make/buy recommendations
- New product involvement

Fig. 7 Changing role of the buyer – from order/expediter to inventory/availability management

Control of inventory levels, on a part-by-part basis, is the responsibility of the buyers. Each buyer has a small number of 'A' class parts to manage as well as a share of the 'B' and 'C' class parts. Each individual has a tight objective in terms of the inventory level of those parts and the total gives one month's supply overall. Each individual is also measured against his/her inventory objective on a monthly basis.

The changing role of the buyer

The role of the buyer in HP has changed significantly over recent years. In fact a more apt job title would be 'professional parts manager' instead of 'buyer'. Instead of simply being an order placer and expediter, the role is now more one of supplier sourcing/ performance measurement/ evaluation, inventory management, negotiation and parts availability management. Consequently, the person specification has changed in terms of professional qualifications and experience (see Figs. 7 and 8).

JIT inventory control has caused a shift in effort from control of a complex works order/tracking system to control of the supplier, the level of inventory and the availability of parts to production. The effort required shifts from the production control department to the purchasing department.

In summary, the successes with JIT at Hewlett-Packard would not have been realised without considerable attention to supplier resource management – selection, communicating expectations and performance measurement/feedback, improved internal and external communication and the organisational and job responsibility changes to support these two areas.

- Business graduate (or equivalent)
- Analytical skills
- Interpersonal skills
- Leadership skills – achieving through others
- Innovative

Part load/work load must be reasonable

Fig. 8 Changing role of the buyer – new set of skills required

6

JIT IN ACTION
CASE STUDIES

The initial application of JIT was in Japan; this was followed by the 'export' of the techniques to Japanese subsidiaries abroad. One subsidiary in particular was the Kawasaki plant in Lincoln Nebraska in the USA. This plant was extensively studied by Richard Schonberger and Robert Hall, whose books based on their respective studies launched the spread of JIT in the West. JIT has been adopted initially by the automotive, automotive supply and the electronics industries. In many cases the initial users were medium and large companies. However, the experience gained by these companies can be translated into other manufacturing sectors and to smaller companies. This section contains both descriptions of JIT in Japanese companies in the West and cases from a wide variety of Western companies who have adopted JIT. This sample covers a representative range of industry, company size and country. They show how JIT has been applied in a wide variety of environments. They illustrate various implementation approaches, and in many cases demonstrate how JIT is being used to support both a company's manufacturing strategy and its strategic market requirements. Finally, they demonstrate the considerable and wide-ranging benefits that can and have been achieved.

Lucas Electrical

J.D. Wood
Lucas Electrical Ltd, UK

This case study covers the period from March 1985 to April 1986 and identifies the primary activities undertaken to arrive at a business redesign in the Lucas Instrumentation business at Ystradgynlais in South Wales.

The activities can be set into four phases (Fig. 1):

- Phase 1 – Business analysis.
- Phase 2 – Detailed analysis of the then current operation.
- Phase 3 – Business redesign.
- Phase 4 – Implementation.

To undertake the task of business redesign it will be appreciated that detailed analysis of all aspects of the business is necessary and indeed not all areas of change and development can be covered within the paper. Elements such as supplier development, control system development, etc., although important in relation to the overall activity, are only mentioned briefly and probably command separate papers in their own right.

Benefits achieved to date as a result of the redesign are as follows:

- Business within a business.
- Product integration.
- Inventory reduction of 60%.
- Lead-time reduction – five days to five hours.
- Improved product quality.
- Productivity increased by 35%.
- Decentralised structure.
- Reduction in indirect manning.
- Increased flexibility – facilities and people.
- Workforce committed to work with change.

Phase 1 – Business analysis

Analysis indicated a broad European customer base and a product variety base of 2500 different units. It was recognised that the products

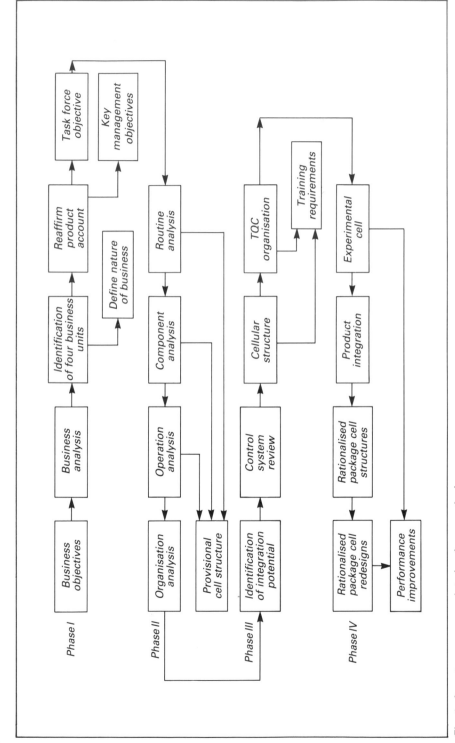

Fig. 1 A systems approach to business redesign

fitted into three families, each having its own characteristics:

- Instrument panels – high volume, low variety, frequently scheduled.
- Cased instruments – low volume, very high variety, generally infrequently scheduled.
- Sensors and transducers – high and low volume.

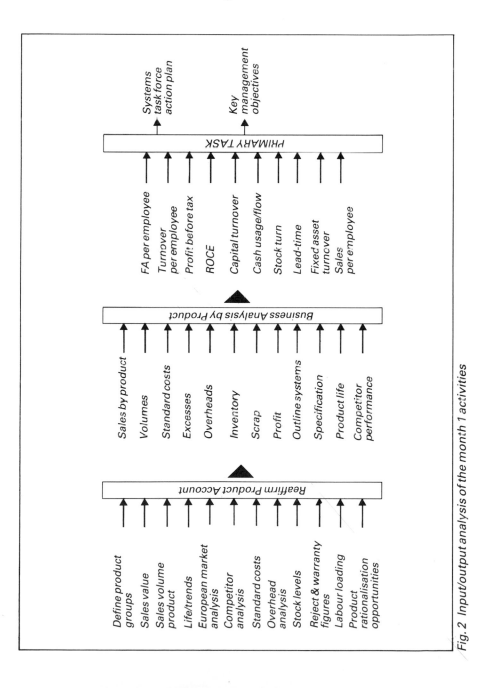

Fig. 2 Input/output analysis of the month 1 activities

Fig. 2 outlines the detail of analysis undertaken and not only relates to the current marketplace, but through a SWOT analysis gave product viability and market share growth opportunities over a five-year period. Competitor analysis supported the SWOT and allowed the marketing strategy for the business to be qualified. An understanding of product life and trends in vehicle instrumentation systems gave the basis for development of the manufacturing strategy.

By reaffirming the product accounts and the families into which a product fitted, a business analysis by product group was undertaken, each being measured against key ratios, resulting in identification of those areas which through a systems task force approach would benefit the overall business's financial performance. From the detailed analysis, other key objectives were identified which would further benefit the business and these were pursued and implemented by the site management.

Phases 2 and 3 – Analysis and redesign

As previously discussed, three businesses (instrument panels, cased instruments, and sensors and transducers) were identified within the business. At operational level, however, each of the business units is supported internally by an electronics manufacturing unit (pcb

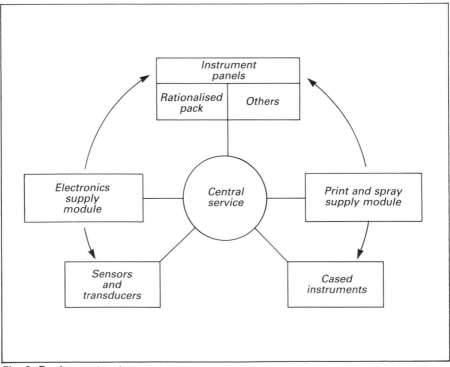

Fig. 3 Business structure

assembly, thick-film substrate manufacture and population) and a print and spray unit; these two were classified as the supply business and operated on a supplier-customer relationship similar to other external suppliers. The overall business structure can be seen in Fig. 3 and shows the traditional central service with which manufacturing industry is familiar; this central service is discussed later in the paper.

A multi-disciplinary task force was formed to analyse the business, and consisted of skills as diverse as business analyst, quality supervisor and work study engineer. This task force of 12 full-time representatives quickly expanded to between 20 and 40 full-time personnel as the then current operation was analysed and moved into the redesign. Business analysis identified the instrument panel business as the area for detailed redesign – this business representing 50% of turnover and good market potential. Specific attention was given to the rationalised instrument panels being produced in volumes of up to 8000 units per week.

It was apparent that manufacture by product was fragmented across the three factory areas and a routing (work flow) analysis (Fig. 4a) shows the excessively high work movement. From the work flow, an activity analysis was produced which identified the relationship of value-added activity to non-value-added. Transportation and delays which add cost, but no value, predominated and reorganisation of the manufacturing facilities was obviously necessary to integrate the product manufacture into the four business units discussed previously. This fragmented operation if integrated would yield the following immediate benefits:

- People would have definitive product areas for which they were responsible and, therefore, accountable.
- Improved communications.
- Improved product quality.
- Less work in process.
- More effective use of people and skills.
- Reductions in work handling and the number of non-value-added activities.
- Problems could be more easily identified and, therefore, resolved quicker.
- Allowed teamwork to be developed.

Component analysis

It was necessary to examine the structure of the bill of materials (BOM) and relate this to the manufacturing procedures in use. It was found that generally Lucas dedicated its products early in the manufacturing routine. Simple reshaping of the procedures pushed the operations that dedicated product by type further towards the end of the manufacturing cycle, which gave clear performance gains and some early definition of how to reorganise people structures.

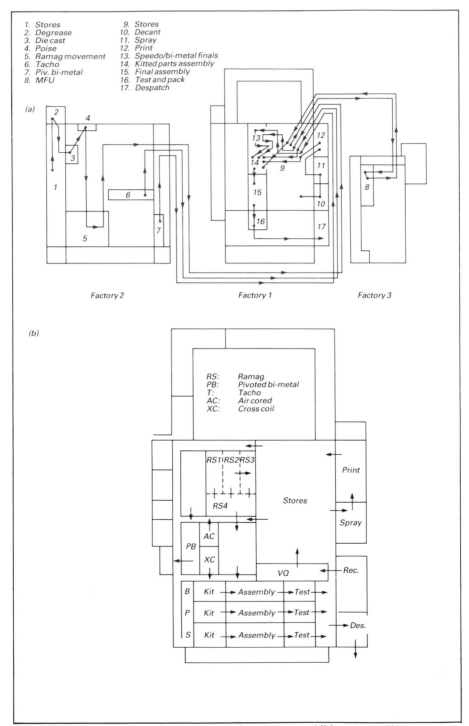

1. Stores
2. Degrease
3. Die cast
4. Poise
5. Ramag movement
6. Tacho
7. Piv. bi-metal
8. MFU
9. Stores
10. Decant
11. Spray
12. Print
13. Speedo/bi-metal finals
14. Kitted parts assembly
15. Final assembly
16. Test and pack
17. Despatch

(a)

Factory 2   Factory 1   Factory 3

(b)

RS: Ramag
PB: Pivoted bi-metal
T: Tacho
AC: Air cored
XC: Cross coil

Fig. 4 (a) Simplified rat pack instrument flowchart, and (b) proposed layout Factory 1

Operational analysis

Each element of the manufacturing routine was examined in detail and a workstation audit produced. The elements examined are shown in Fig. 5 and allowed new tooling, equipment handling facilities, etc. to be identified and ordered to meet the redesign criteria.

Lead-time/response time analysis

In conducting the detailed analysis of the operation, lead- and response times were identified in days and in some areas weeks, and became a criterion for improvement in the redesign.

Organisational structure

Recognised as functional and hierarchical, Lucas operated with central control and not operational control. This gave rise to an individualistic approach with vague responsibility and accountability. Problems occurring in manufacture could not be resolved fast enough and, indeed, due to the high raw material availability in some instances, never were.

Reorganisation of people structures was necessary and the objectives clear – devolve those skills necessary in support of operational improvement into the business units, to create a 'cellular' operation embracing teamwork, flexible working and continuous improvement.

A number of cells would form a module and a number of modules

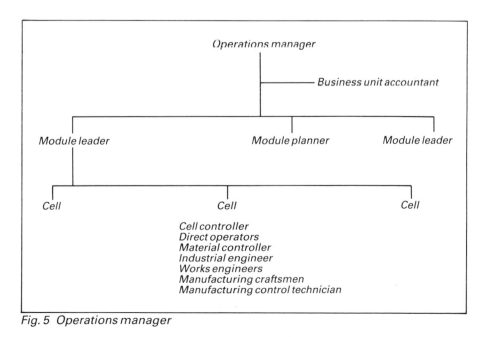

Fig. 5 Operations manager

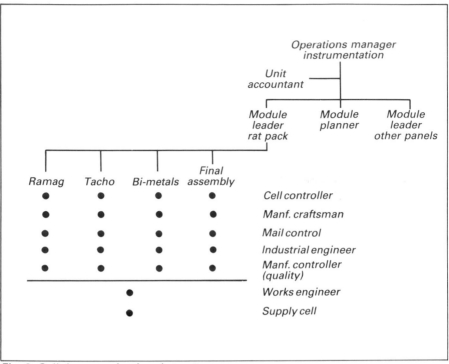

Fig. 6 Cellular organisational structure

constitute a business. Fig. 6 shows the general cell structure and a business structure.

The detailed business analysis and the instrument panel redesign benefits, allowed development of a total business plan to achieve both change and the significant benefits that would result from it. The redesign of the instrument panel business is shown in Fig. 4b. The operating improvements and benefits are summarised below:

- Responsibility for business performance can be identified at business level.
- Kanban could be introduced.
- Integrated manufacture.
- Reduction in non-value-added activity.
- Improved control.
- Buy-off.
- Reduction in indirect support.
- Productivity improvement of 25%.
- Inventory reduction of 60%.
- Stockturn improvement to 36.
- Lead-time improvement five days to five hours.

With the level of improvement identified, it was necessary to simulate the redesign to qualify and optimise these benefits. This was attempted manually initially and although Lucas could test the design statically, it was impossible to monitor under dynamics of change. The use of the Hocus simulation package from P-E Consulting Services, however, allowed dynamic testing and optimisation of cell, module and business performance. The Hocus package became an even more important tool when implementation commenced, as it allowed sensitivity measurement to take place with true shop-floor dynamics; obviously this work still continues, allowing further gains to be achieved in cell performance.

Having qualified the bottom-up design, work commenced on identifying the bottom-up and top-down control system needs of the business devolving the strategy and plans for system integration, development and implementation. Having identified the people structure to support the bottom-up design and formalising the necessary changes in working practices to achieve labour flexibility and team work, Lucas could begin to re-examine the functional structures within the business. Development of skills and changed responsibilities within cells reduced the external support service requirements and by using ZBA, reorganisation of the support services was achieved allowing change to a TQC structure to be achieved over a two-year period.

Identifying the characteristics of the business redesign brought to the surface certain key activities which would be required to be developed to enable the business to move towards a true 'Just-in-time' development plan. With lead- and response time reductions being so dramatic within the redesign concepts, supplier and customer development programmes had to be developed to support the JIT strategy.

It will be recognised that change of this magnitude necessitated the development of training plans for all levels of personnel within the business.

Phase 4 – Implementation

The key features of this phase include: product integration, product rationalisation, experimental cell, supplier development, customer development, simulation, control system (top-down/bottom-up), payment scheme review, and training.

Product integration

Integration commenced in August 1985 and was completed by November 1985. The instrument panel business was sited in Factory 1; cased instruments, sensors and transducers in Factory 2; and the supply module in Factory 3.

Immediate benefits were achieved: reduced transportation, storage and delays, improved quality, reduced WIP, better communications, floor-space saving and easier identification of problems.

Experimental cell

Recognising the nature and order of change proposed for the business, it was thought necessary to set up an experimental cell where Kanban, SPC, problem-solving skills and teamwork could be demonstrated.

The speed transducer product group was chosen for the experiment and the objectives defined:

- Develop awareness.
- Demonstrate new job roles and the team approach.
- Marketing vehicle – to sell the approach within the business.
- Allow feedback – shop-floor and managerial.

Agreement was reached with the unions that within the cell all personnel would operate flexibly and develop practices as required for a period of six months.

Key features of the speed transducer are: product variety, 22; standard runners, 6; strangers, 16; and BOM levels, 8.

The cell was redesigned to work with Kanban and optimised using Hocus simulation. The area was reconstructed, throwing out 'work storage' conveyors and the now classic U-shaped line introduced, supported by sub-cell operations. Raw material stores were brought on line, piecework removed and operators trained on Kanban and SPC.

Although a simple tool to use, Kanban does demand rigid disciplines by all who work with it to sustain the high levels of control it offers. The Lucas-developed Kanban training package was used to train all cell personnel in Kanban, and manufacturing systems engineers actually train the team and indeed join the team when the system goes 'live'. The systems engineers support the teams for approximately six months, guiding, encouraging and policing the maintenance of Kanban discipline.

SPC is introduced in a similar manner except training is conducted by the training and quality departments.

The cell was structured as described previously (Fig. 6), with key skills and resources being introduced to ensure quick responses and resolution to problems that occurred. The key elements, teamwork and continuous improvement, were not easy to introduce; and many lessons were learnt and most certainly mistakes made. The most important lesson was that management *cannot* say 'yes' and walk away; their involvement, participation and support is mandatory unless you want to fail.

It is interesting to note that although initially everyone claimed to wish to participate in a team approach, in practice it was found that at first some were only prepared to go part of the way. Although it is necessary to emphasise the need to find the right blend of skill, attitude becomes the predominator in achieving a successful team. The perfect blend of skill and attitude is rarely found, but one can recognise that those individuals with real commitment to make teamwork and change happen, sway the dissenters into following the team approach.

There was surprise at the effectiveness of Kanban, being way above perceived understanding. Bottlenecks and scrap generators were clearly identified, known to the whole team and required immediate solution. With the cell structure designed to incorporate additional skill levels, the response to problems and the solving of them was much enhanced. However, more could be achieved and continuous improvement (quality circle) activity was developed.

The results were astounding, not only in business performance terms but also in enhancing the commitment of people. They quickly adopted the business as theirs, and it was of course. They realised they could be involved, play a real part and influence the performance of 'their business'. This team, this experiment achieved:

- Operational in mid-November 1985.
- Work in progress reduced by 84% in four weeks.
- Raw material reduction of 40%.
- Productivity improvement of 22%.
- Stockturn to 15.
- Lead-time reduction.
- Improved product quality.
- Changeover time reduction from 30 to 5 min.

Of course this was only an experiment, but all the obejctives were met. On 22 February 1986 agreement was reached with the unions to proceed with the implementation of change. The agreement incorporated flexible working, removal of restrictive practices and lines of demarcation, and the continued development of 'the team approach'.

Following the success of the experimental cell, further cells were implemented; it was decided that although the whole of the instrument panel business was to be redesigned using Kanban, it would be advantageous to redesign a cell within the cased instrument business, thereby having change operational in each major business unit. The electronic speedometer line was the cell chosen and the changes in layout and benefits achieved are shown in Fig. 7. Again, one can see the U-shaped facility in the redesign; the old conveyor line being unnecessary. It is interesting to note that in the four months of redesign and implementation, 100 metres of conveyor were disposed of.

Instrument panel redesign and implementation

The overall benefits of redesign in the panel business have already been discussed. Implementation has commenced, with final assembly being achieved from a kitting station where all non-standard elements of the 24 varieties built are placed into a box and barcode labelled. Assembly is by unit build and work delivered by a computer controlled conveyor,

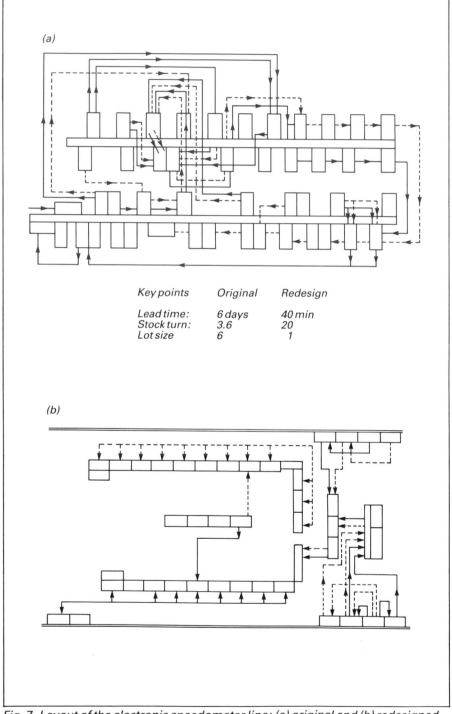

(a)

Key points	Original	Redesign
Lead time:	6 days	40 min
Stock turn:	3.6	20
Lot size	6	1

(b)

Fig. 7 Layout of the electronic speedometer line: (a) original and (b) redesigned using Kanban

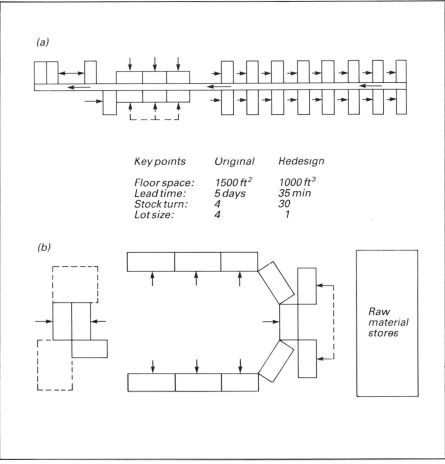

Fig. 8 Instrument panel redesign: (a) original tacho cell, and (b) following Kanban redesign

allowing a 'one-on-one' build capability and WIP control.

A diagrammatic implementation is shown in Fig. 8. This is a traditional facility with conveyor and where high WIP was replaced with a U-shaped line operating on the concept of 'kanban squares'. An empty square provides the trigger to build, allowing only one unit between operators.

Key features of this line include: batch size of one, kanban squares, WIP reduction of 99%, total WIP on lines of 14 units, lead-time reduction from five days to 35min, raw material reduction of 40%, floor-space reduction of 30%, elimination of conveyors, and ratio of value-added to non-value-added increased by 60%.

Total redesign of instrument panels was programmed for completion in June 1986.

Concluding remarks

This paper demonstrates a structured approach to business redesign and, although not all facets have been covered in the paper, the detailed analysis required to support the activities undertaken is obvious. The approach to Just-in-time manufacturing embraces all of the elements highlighted; it is not simply building to order or having materials available to build, it is much deeper if a real JIT approach is to be adopted. The paper shows the methodology being adopted across Lucas Industries in the pursuance of manufacturing excellence, but let us not forget that no matter how sound our concepts and technologies are, success depends on management developing and training the workforce of industry and stimulating its involvement in the performance and effectiveness of their operations.

Efforts are continuing in Lucas Instrumentation to still further improve the business performance, with a workforce totally committed to work with change and continuous improvement. In conclusion, some of the benefits and achievements attained since March 1985 include:

- A workforce commited to work with change.
- Flexibility in people and facilities, no longer stock.
- Business within business.
- Decentralised structure.
- Simple control systems.
- Reductions in indirect manning.
- Teamwork and continuous improvement.
- Kanban implemented.
- Improved communication and problem solving.
- Supplier/customer development operating.
- Improved product quality.
- Inventory reduction of 60%.
- Lead-times reduced.
- Productivity improvement of 30% across the business.
- Improved business financial performance.

Rank Xerox

B. Zimmermann
Rank Xerox Manufacturing BV, The Netherlands

Investigations into automatic material handling capabilities have resulted in the development and the introduction of the 'Automatic Material Control System' (AMACS) in Rank Xerox Venray. This system controls all material movements from receipt up to the point of assembly, except for the storage of non-Just-in-time material in the conventional stores.

The main objectives of this system are to comply with the Just-in-time strategy, and to realise cost savings in the field of material flow control.

This paper gives an outline of the logistic considerations and describes the configuration and functions of the system.

Logistic strategy

The strategy is as follows:

- To receive material just in time for production – This can be realised by providing parts under a 'quality' certificate. Preferably these will be expensive and voluminous parts.
- To buffer JIT material directly available to production – Not in the conventional stores, where direct issue to production is not immediately possible, but rather in an automated JIT buffer as work in process, allocated to work orders prior to or upon receipt.
- To issue and deliver JIT material immediately to production – Upon request of each individual workstation in the production, material will be delivered to the point of assembly. This pull principle also applies to conventionally stored non-JIT material.

Requirements profile for the Just-in-time material flow

For the JIT strategy to be effective, the following items have to be considered and specified criteria must be achieved:

- Quality certification.
- Vendor packaging.

- Transport coordination.
- Just-in-time buffer.
- Transport system to production.

A more detailed explanation is given in the following sections. A block diagram of the JIT material flow is given in Fig. 1.

'Quality' certified material

Just-in-time material must be certified material. This quality status is achieved by certifying the supplier's manufacturing process, and training the supplier's personnel in the application of Rank Xerox quality standards and procedures for the certification of parts.

Vendor packaging

In order to avoid delay in material replenishment due to repack, JIT

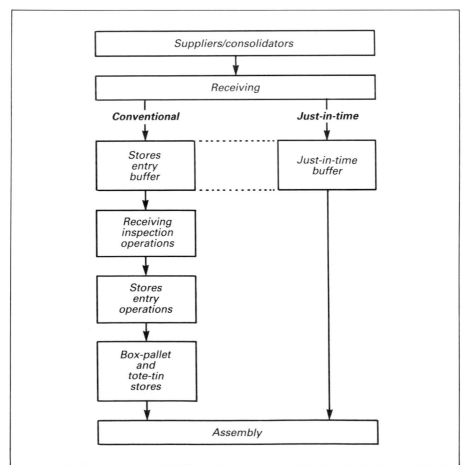

Fig. 1 Just-in-time material flow block diagram

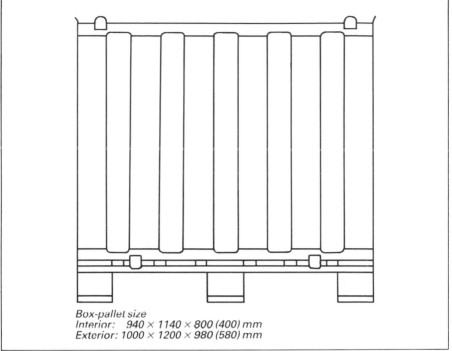

Box-pallet size
Interior: 940 × 1140 × 800 (400) mm
Exterior: 1000 × 1200 × 980 (580) mm

Fig. 2 Box-pallet (front view)

parts must be packed according to Rank Xerox, quality protective, vendor packaging instructions on standard four-way entry pallets, and on standard four-way entry box-pallets (Fig. 2).

Non-JIT items must also be packed according to Rank Xerox vendor packaging instructions either in/on the standard box-pallet or pallet, and in/on the standard tote-tin (Fig. 3) of which up to 20 can be stacked onto a pallet. Where economically justified, Rank Xerox Venray standard containers will be delivered to the suppliers.

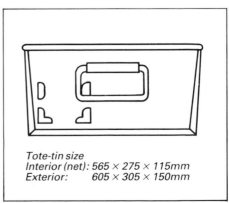

Tote-tin size
Interior (net): 565 × 275 × 115mm
Exterior: 605 × 305 × 150mm

Fig. 3 Tote-tin (front view)

In cases where distance and volume do not permit application of standard containers, the suppliers pack according to the Rank Xerox 'General Supplier Packaging Specification' which mainly prescribes:

- The standard four-way entry pallet.
- A strapped boxing suitable for easy re-toting and optimised cube utilisation.
- Quality protection.
- Identification.

Transport control and consolidation

The Just-in-time strategy tends to decrease the batch sizes and therefore increases the transport costs due to *higher delivery frequencies*. This disadvantage must be overcome by a transport consolidation system which is interfaced with the plant's material requirement forecast.

The Just-in-time buffer

The JIT buffer must interact with the goods receiving function at the docks, and the assembly workstation. It therefore communicates with the transport system operating between JIT buffer, stores, and each individual workstation. The JIT buffer must provide the additional facility to hold and issue non-JIT parts prior to quality checks and conventional stores entry process.

Transport system between JIT buffer/stores and workstations

This transport system must be capable of:

- Operating at the individual workstation's call for parts replenishment from the JIT buffer and conventional store.
- Interacting with the JIT buffer and the conventional store for parts issue.
- Transporting subassemblies from the sublines to the mainlines.
- Transporting finished products to the shipping dock.
- Transporting empty containers from the workstations for recycling or other destinations.

Material transport and storage equipment

Receiving area

Material arrives at two receiving points (Fig. 4):

- The primary area having four docks and accepting loads from the transport consolidation and directly from suppliers.
- The secondary area is a single-point receival draughtgate to accept onsite material transfers into the building.

All material arrives at the receival points on standard four-way entry

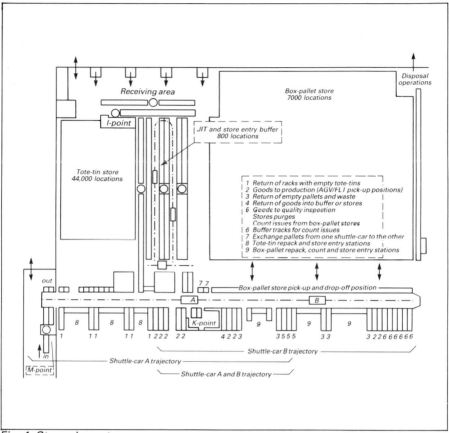

Fig. 4 Stores layout

pallets of 1000 × 1200mm dimensions. Pallets are unloaded by manned fork-lift trucks (Fig. 5).

In the main receiving area the pallets are deposited onto three track positions. These receival points are designed for two-way directional access, so that pallets can be dropped regardless of direction of pallet dimension.

Rollertrack transportation moves the pallet into an automatic dimension and weight checking area. Failure to comply with the weight and profile criteria triggers rejection of the pallet which is then automatically transported to an area for attention.

The secondary receiving area is also equipped with rollertracks to accept onsite material and has a self-contained weight and profile check facility.

Automated JIT and stores entry buffer

Acceptance of a pallet triggers transportation to the centre of the JIT and stores entry buffer following keyboard operator interaction with

Fig. 5 Pallets are unloaded by manned fork-lift trucks. In the main receiving area the pallets are deposited onto three track positions. These receival points are designed for two-way directional access, so that pallets can be dropped regardless of direction of pallet dimensions

the pallet data and the crane and shuttle-car computer (Fig. 6). In fact the goods inward buffer is a relatively small, fully automated pallet store with a capacity of 800 pallets.

Pallets for buffering will be located by computer controlled execution

Fig. 6 Acceptance of a pallet triggers transportation to the centre of the JIT and stores entry buffer following keyboard operator interaction with the pallet data and the crane and shuttle-car computer. The pallets are handled by two computer-controlled autocranes

Fig. 7 A rollertrack-cart distributes the pallets from the exit tracks

into a location as close as possible to the next location to be issued. The pallets are handled by two computer controlled autocranes. Incoming pallets from both receival points are transported via rollertracks to transfer turntables in the centre of the buffer which move the pallets into a pick-up position for the autocranes. Between these two pick-up points is one exit transfer turntable onto which the cranes unload the outgoing pallets. Via this turntable the pallets are moved onto two exit rollertracks and transported to the shuttle-cars for onwards distribution to the pick-up rollertracks for automated guided vehicles (AGVs).

The exit flow from the buffer is spread as much as possible in order to avoid queues, to employ the capacity of the two twin-forked shuttle-cars, and to optimise the loading of the shuttle-cars. For this purpose a rollertrack cart distributes the pallets from the exit tracks over two tracks for the tote-tin store, two for the box-pallet store, and two tracks for production (Fig. 7).

Shuttle-cars

A shuttle-car is a twin-forked automatic vehicle. Each pair of forks can handle a pallet independent of the other (Fig. 8). Two shuttle-cars operating on rails are the connecting links between the buffer store output tracks and the final destinations of the pallets (Fig. 9). In fact these vehicles take care of all material movements as shown in Fig. 10.

Along the railtrack there are 112 drop-off and pick-up positions. To meet the capacity demands, the movements of the two shuttle-cars are optimised as much as possible. There is one order list for both cars and all pick-up and drop-off positions are installed on a grid to allow for

Fig. 8 Each pair of forks can handle a pallet independent of the other

Fig. 9 Two on-rails operating shuttle-cars are the connecting links between the buffer store output tracks and the final destinations of the pallets. In fact these vehicles take care of all material movements between the parallel to the rail trajectory situated areas and activities

FROM	↓ →	a	b	c	d	TO e	f	g	h	j
Buffer	a	–	+	+	–	–	–	+	+	–
Box-pallet repack stations	b	–	–	–	+	–	+	+	+	–
Tote-tin repack stations	c	–	–	–	–	+	–	+	+	–
Box-pallet store	d	–	+	–	–	–	–	+	+	–
Tote-tin store	e	–	–	–	–	–	+	+	+	–
Draughtgate	f	+	+	+	–	+	–	+	–	–
K-point rollertracks	g	+	+	+	+	+	+	–	+	–
AGV/FLT transport	h	–	+	+	–	–	–	+	–	+
Count-issue buffer	j	–	+	–	–	–	–	–	–	–

Fig. 10 Shuttle-car interactions

double pallet handling (Fig. 11). In the centre of the track is an overlap area where both shuttle-cars can operate.

AGV system

The control system. The total integrated control system consists of the following major components:

- Carriers with on-board Zilog 80 microprocessor which controls the vehicle.
- The control loops and frequency generators. The floor layout is connected to the frequency generators which provide 2–5kHz alternating current for the loops. The carriers sense the magnetic

Fig. 11 All pick-up and drop-off positions are installed on a grid to allow for double pallet handling

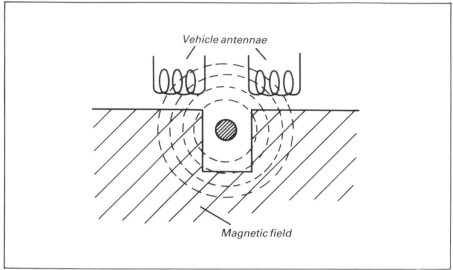

Fig. 12 Vehicle antennae

field from the inductive control loops (Fig. 12) and follow the route as programmed in the on-board microprocessor. This guiding principle is used along the mainlines where the carriers follow one of the four different frequencies dependent upon the instructions received from the control computer when operating in automatic mode, or through the on-board control panel when operating in manual mode.

● The control computer is a PDP 11/23+ and has the following primary functions:
  – receipt of transport orders through interface with AMACS or from a terminal
  – generating transport assignments to carriers
  – control of individual carrier movements and traffic control
  – exchange of information between carriers and PDP
  – carrier routing and order assignments
  – status reporting to AMACS
● Radio communication (450MHz) is used between the carriers and the PDP 11/23+ to achieve duplex data transmission.

The carriers. The AGV is a fork-type vehicle encased in stainless steel. It is powered by two batteries sufficient for 16-hour's operation. In any case the charging status is automatically checked after completion of a mission. Depending on the status of the batteries, the control computer decides whether the AGV can take up the next assignment or be sent to a recharge station.

Battery charging is carried out over 8 hours (non-production time) in an area which permits each carrier to enter and leave without

Fig. 13 Battery charging will be carried out over 8 hours (non-production time) in an area which will permit each carrier to enter and leave without interfering with other carriers

interfering with other carriers (Fig. 13).

Safety and anti-collision protection of the vehicle and its load is achieved by the following (Fig. 14):

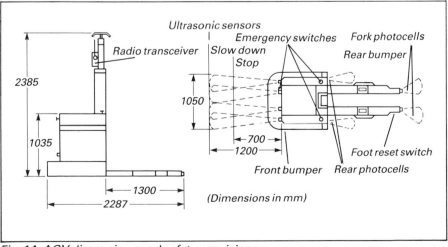

Fig. 14 AGV dimensions and safety provisions

- Front bumper – Any movement of the bumper causes limit switches to interrupt the travel and lifting movements of the truck.
- Rear bumpers – Limiting switches are incorporated into the forks which are actuated by depressing spring blocks on the tips of the forks.
- Fork photocells – Photocells in the fork tips with a restricted field of visibility can be program-controlled to enable fork travel under a load.
- Rear photocells – Photocells on each side of the vehicle facing the rear, to avoid risk of jamming accidents occurring between the truck body and an obstacle. The photocell stops reverse travel if sensed.
- Ultrasonic sensors – These sensors are at the front of the vehicle, and slow it down when activated in conditions such as narrow aisles or approaching obstacles. They stop the carrier when the distance between an obstacle and carrier becomes less than 700mm.
- Foot reset switch – Each vehicle is fitted with a foot-operated reset switch. This is used when a vehicle has been waiting for an operator to change over tote-tins, and is the instruction for the vehicle to continue.
- Emergency switches – There are three emergency switches installed which when activated immediately stop all movements of the vehicle.

The carriers have three modes of operation:

- Automatic control by the PDP computer.
- Semi-automatic control via a keyboard.
- Manual control by a hand-held controller on board the vehicle.

Layout details. The total floor wiring layout contains 1710 locations for delivery or pick-up of material, of which 1650 are in the production area and 60 in the stores area. The routing of the vehicles is based on one-way traffic between the production lines and two-directional traffic in the main supply aisles as shown in Fig. 15.

AGV locations can be deleted or reactivated as required. Each location is 1150mm wide, thereby leaving a clearance of 150mm between two pallets. All locations are aligned to the left and right of the induction wire.

Where locations are allocated to delivery of tote-tins, the AGV will penetrate a further 400mm to bring the tote-tins closer to the operator.

Installation of floor wiring. All lines are measured and marked on the floor using laser beam equipment. These were 'cut' using diamond tools. The floor and surroundings were kept dust-free by using a waterspray when cutting took place.

The wires inserted in the grooves are firstly covered with a rubber strip, followed by an epoxy resin which is grinded if additional

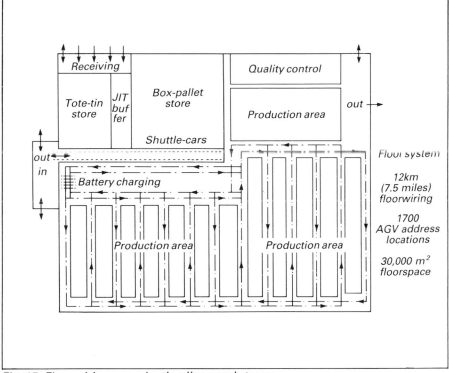

Fig. 15 Floorwiring – production lines and stores

smoothness of the floor is necessary.

System configuration

The Automatic Material Control System (AMACS) is responsible for controlling material movements to and from stores, processing its arrival and coordinating its dispatch to workstation requests.

There are various computer systems contributing to the automatic material control system (Figs. 16 and 17):

- *AGV process computer* – controls the inductive guided vehicles in the assembly area.

- *Auto crane and shuttle-car process computer* – controls the two autocranes in the buffer store, the two shuttle-cars and the various rollertracks including the pick-up points.

- *Assembly terminal network system* – acts as a concentrator for the alphanumeric terminals at the production workstation.

- *XBMS (XCS applications)* – this is the material requirement planning system handling orders and performing additional business managing functions.

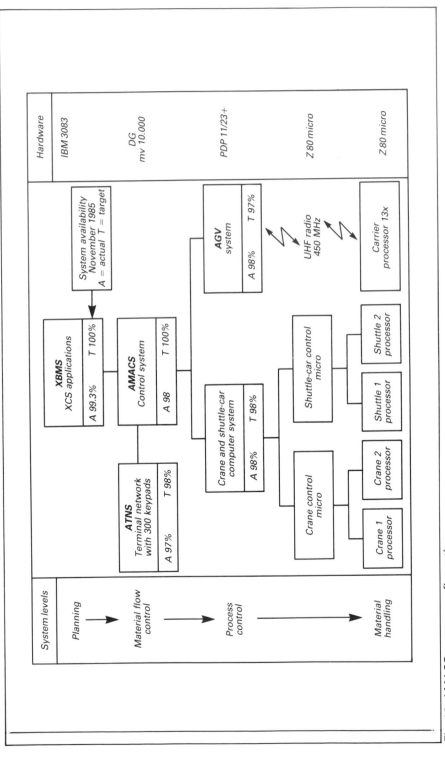

Fig. 16 AMACS system configuration

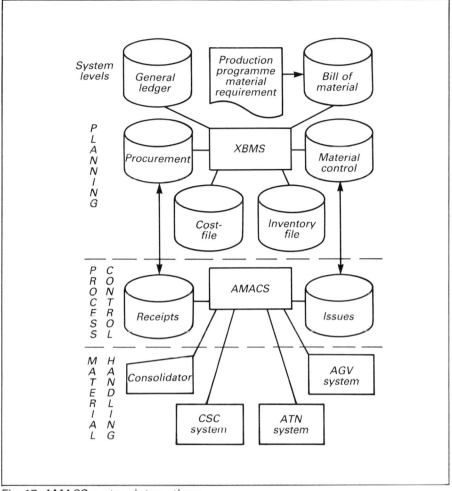

Fig. 17 AMACS system interactions

Daily throughput

AMACS supports processing associated with the following daily throughput while the system is on-line:

- 640 receipts (lot information).
- 1420 pallet arrivals (pallet data from crane and shuttle-car process computer).
- 790 pallets dispatches to repack stations.
- 4220 requests from workstations.

Physical material handling limitations mean that the above load is distributed evenly throughout a 16-hour day of issuing and 12 hours of receiving. When the system is off-line, it is able to support 3400 picking tickets arriving from XBMS.

Capacity

The AMACS database is capable of storing information for all of the following logical entities:

- 1500 lots. A lot is established by the execution of a receipt. Lot information is deleted when the number of associated pallets in the goods inward buffer, or in transit to or from the buffer, falls to zero.
- 1500 pallets. A pallet becomes known to AMACS when the crane and shuttle-car process computer transmits a pallet data message.
- 15,000 picking tickets.
- 2000 workstation requests. A workstation request is established when AMACS processes a request code. It ceases to exist when AMACS completes the necessary deliveries.
- 5000 workstation request codes.
- 3000 part/vendor relationships identifying the combination of part and vendor for which lots are processed as JIT lots. The relationship also provides information on quantity currently available for allocation, as well as re-order quantity level and source.
- 16,000 parts.
- 100 tote-tin batches. A batch exists from the time at which the first picking ticket is associated with it until the last tote-tin has been delivered.
- 100 shortages. A shortage condition arises when a workstation request cannot be satisfied. The shortage condition disappears when the suitable material is made available for allocation.
- 14 days of history data to support the printing of history reports.

Functional system description

Receipt processing

The AMACS database maintains details about the different types of parts known to the system. One important item of information about each part is its packaging type. AMACS recognises three classes of part; namely, tote-tin parts, box-pallet parts, and JIT parts.

As far as the AMACS system is concerned, parts arrive in lots (a lot is a collection of identical parts). A single lot may be distributed over multiple pallets and a single pallet may contain parts from several lots. However, a JIT part may not be on a pallet together with another JIT or non-JIT part.

When a collection of pallets arrives at the receiving dock from an external supplier, an operator interacts with an AMACS terminal in order to execute a receipt transaction, in which details of the various lots in the delivery are entered.

AMACS passes details of each lot to the XBMS system (AMRA trancode), which is awaiting the arrival of the lot, having been responsible for ordering it from the supplier.

XBMS assigns a lot number for use by AMACS in tracking the parts in the lot, and prints a goods movement slip, which stores personnel attach to the first pallet of the lot.

If XBMS is currently not available, AMACS allocates the lot number locally. The information which would ahve been transmitted to XBMS is stored (along with the lot numbers allocated by AMACS) until XBMS becomes available, at which time it is transmitted as a batch. AMACS also passes details to the crane and shuttle-car process computer.

Arrivals from the main consolidators are handled in the same way as those direct from the suppliers, except that the receipt transaction is initiated by an operator at the main consolidator warehouse prior to dispatch, rather than when the lots arrive at the receiving dock. Goods arriving at the receiving draughtgate are internally manufactured parts.

AMACS treats such lots in a similar way as any other lot within the system except that there is no interaction with XBMS.

Receipt buffer

Once the receipt transaction has been executed and the parts have passed one of the receiving areas, they enter the JIT and stores entry buffer. Before this can happen, each pallet in the delivery has to be identified to the system, together with its contents. The crane and shuttle-car process computer generates and transmits to AMACS a pallet number, along with information about which lots are represented on the pallet and the number of parts from each lot present.

Repack/stores entry operations

Pallets arriving from the receiving docks or the receiving draughtgate are temporarily stored in the JIT and stores entry buffer under the control of the crane and shuttle-car process computer.

Pallets containing JIT parts remain in the buffer until dispatched in response to assembly workstation requests.

Non-JIT parts are passed on to repack stations as soon as possible after arrival. The allocation of pallets to the repack stations is under control of the crane and shuttle-car computer.

Stores personnel are responsible for repacking/retoting operations and sample preparation for quality inspection.

Finally non-JIT parts are stored in a conventional (i.e. non-automatic) pallet or tote-tin store.

The assembly area

Equipment is manufactured in an assembly area which contains 24 production lines (Fig. 15). Each production line occupies a long, relatively narrow, strip of floorspace. This strip itself is divided into three adjacent parallel strips (Fig. 18). One is occupied by the line which transports the partially assembled equipment. The next is

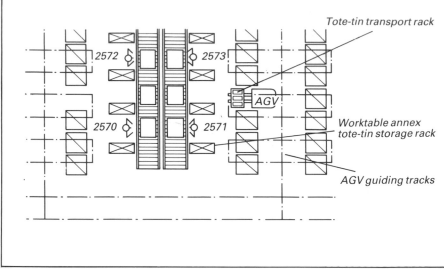

Fig. 18 Assembly workstations

occupied by the assembly workers, their equipment and some of their individual stocks of parts for the assembly process. These parts are normally small parts held in metal boxes called tote-tins. The third strip holds additional parts, carried on pallets. This strip is divided along its length into locations, each large enough to hold a single pallet.

The production line is divided into workstations. A workstation is that part of the line at which a particular part of the assembly process is performed. It is manned by an assembly worker and includes a number of adjacent pallet locations.

Between the production lines run tracks for AGVs which can transport material and packaging to and from the AGV locations. The tracks are organised into a number of AGV loops. An AGV loop serves a fixed set of production lines.

Certain production lines are served by fork-lift trucks rather than by AGVs. The drivers receive printed instructions telling them what to transport and where to deliver it.

Several different types of part may be held at a workstation, each in its own dedicated storage position. The parts are held in one of two ways: on a pallet or in a tote-tin. When an assembly worker wishes his stock to be replenished he enters a code on an alphanumeric terminal (Fig. 19). Some time later his request is served by one or more AGVs. If the part is stored on a pallet, an AGV collects the empty pallet, after which it (or another AGV) delivers a pallet containing the requested part. If the part is stored in a tote-tin, an AGV arrives with a transport track filled with tote-tins. The operator then exchanges his empty tin for a full one, being careful to select the correct tin from the rack (Fig. 20). In order to receive tote-tins, a workstation must have an unoccupied

Fig. 19 When a workstation operator requires his stock of a particular part to be replenished, he consults his menu card, identifies the appropriate code and enters it on the keyboard of a small terminal associated with the workstation

AGV location via which the AGV can enter the station with its rack.

A tote-tin contains a stock of a single part. A tote-tin transport rack may contain up to 14 tote-tins. The use of multiple tote-tin transport racks allow stores personnel to batch up deliveries of tote-tin parts so as to economise on AGV movements.

Fig. 20 The operator exchanges his empty tin for a full one, being careful to select the correct tin from the rack

An AGV collects a transport rack from one of the stores exit-tracks and makes a tour through the assembly area, stopping at a specific set of stations. All stations on the tour must belong to the same AGV loop.

Workstation requests

From time to time, changes in production demand and other factors require the distribution of activities between workstations to be modified. Such an operation is known as rebalancing and is carried out with the help of applications in XBMS. Whenever the functions of a workstation are established or modified following rebalancing, XBMS provides AMACS with a list of parts to be used at the workstation.

An AMACS transaction is then used to associate each part with a unique request code and a range of dates during which the code may be used, and to identify the AGV location to which the part is to be delivered for use at the workstation (a single AGV location is allocated as a destination for all tote-tin deliveries for the workstation).

A menu card (Fig. 21) is printed for use by the workstation operator in identifying the request code corresponding to each part. The menu card also identifies certain special codes, which allow pallets to be moved around within the assembly areas or to be returned to the stores.

When a workstation operator requires his stock of a particular part to be replenished, he consults his menu card, identifies the appropriate code and enters it on the keyboard of a small terminal associated with the workstation. AMACS interprets the code as identifying a request for a particular part to be delivered to a particular AGV location, belonging to a particular workstation.

Request code menu for terminal 25   *Printed 01-10-1983*

Station	Code	Part number	Description	AGV	Start	End	Comment
2570	E201	1 S 91678	Registr. ass.	E 04			Subassembly
	BA02	3 S 90821	Hinge ass.	E 04			
	F403	7 S 90499	Idler ass.	E 04			Subassembly
	2C04	11 S 90300	Lever ass.	E 04			
	B305	15 P 02335	Clip	E 04	15-10-	20-11-	
	306C	19 P 90464	Capstan clamp	E 04	83	83	
	A307	20 P 60639	Capstan front	E 03			
	B608	26 P 90753	Sgt screw	E 04			
	209C	26 P 90870	Screw	E 04			
	AB10	30 P 93552	Bracket cable	E 04			
	F311	30 P 93553	Bracket cable	E 02			
	12C5	30 P 94092	Bracket hinge	E 01			
	A135						Refuse disposal

Fig. 21 Workstation request menu

Issues via picking tickets

The administration system XBMS is used for a number of tasks. Production planning applications allow it to predict the requirements for material in the assembly area. Other applications allow stores personnel to keep XBMS informed as to the exact contents of individual locations within the non-automated parts of stores, i.e. the box-pallet and tote-tin areas. As a result, XBMS is able to generate picking tickets.

Picking tickets generated by XBMS are transmitted to AMACS at night (i.e. before manufacturing starts) for the following day. A picking ticket informs AMACS that it can expect a 'pull' request from a workstation at some time during the coming day. When AMACS is attempting to satisfy a workstation request, it tries to match the request to one of its stored picking tickets. A match only occurs when the request and picking ticket identify the same part number *and* next higher assembly. Failure to find a match causes AMACS to attempt to allocate the parts from the JIT buffer.

A picking ticket is printed when AMACS detects a match between the contents of the ticket and a workstation request for parts.

Extraction of parts from stores

The picking ticket identifies the number of parts to be supplied and where they are to be found (a picking location). AMACS prints the picking ticket (which includes a ticket number printed in barcode format) on a printer in the stores area. Stores personnel then manually retrieve the parts from stores. AMACS does not re-enter the process until a barcoded ticket is read at stores exit.

When a barcode is read at one of the exit positions in front of the box-pallet store, AMACS instructs the crane and shuttle-car process computer to deliver the pallet to one of the stores exit tracks. The process computer chooses the exit track based on the station number of the workstation to which the pallet is to be delivered. In case the workstation is in an area which is served by a fork-lift truck the pallet will be taken to an exit track dedicated to fork-lift truck transportation.

The crane and shuttle-car process computer reports the arrival of a pallet at the end of an exit track to AMACS. Depending on the identity of the track AMACS either instructs the AGV process computer to transport the pallet to the appropriate AGV location at the workstation, or prints a delivery instruction for a fork-lift truck driver. AMACS also informs XBMS that the picking ticket has been used. Processing is complete when the AGV process computer has reported that the pallet has been delivered, or immediately if a fork-lift truck is involved. AMACS deletes the picking ticket and re-enables use of the request code at the workstation.

Material deliveries from the tote-tin store are batched-up on transport racks. Successive reads of barcoded picking tickets at one of

Fig. 22 When the transport rack arrives at the end of an exit track, AMACS instructs the AGV process computer to collect the rack and to tour the workstations

the tote-tin store exit points define a tote-tin batch corresponding to the contents of a transport rack. When a rack is full, the operator instructs the crane and shuttle-car process computer to transport the rack to one of the store's exit tracks. When the crane and shuttle-car process computer informs AMACS that the rack has arrived at the end of a store's exit track, AMACS instructs the AGV process computer to collect the rack, to tour the workstations identified in the picking tickets and to return the rack to stores (Fig. 22). No such instruction is issued if the exit track is served by a fork-lift truck. AMACS also informs XBMS that the picking tickets have been used to satisfy requests from workstations. Picking tickets are deleted from the AMACS database one or more at a time, as the AGV process computer reports delivery of tote-tins to specific AGV locations, or immediately if the exit track is served by a fork-lift truck. The corresponding request codes are re-enabled at the same time.

Issues from JIT buffer

If a workstation request cannot be satisfied via the picking-ticket mechanism, AMACS checks its files to determine whether the part can be supplied from the JIT buffer. In the search process, only the part number is significant, AMACS does not keep next-higher-assembly information for parts in the JIT buffer.

If the JIT buffer contains a pallet holding parts of the correct type,

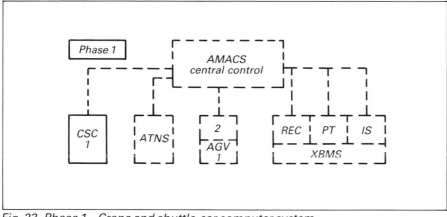

Fig. 23 Phase 1 – Crane and shuttle-car computer system

AMACS instructs the crane and shuttle-car process computer to extract the pallet from the buffer and to transfer it to an appropriate store's exit track.

If the current issue reduces the stock of parts in the JIT store to below a re-order level defined for that part, a message is output on a printer.

When the crane and shuttle-car process computer informs AMACS that the pallet has arrived at the end of a store's exit track, AMACS instructs the AGV process computer to transport the pallet to the correct AGV location. The workstation request is satisfied when the AGV process computer informs AMACS that the pallet has been delivered.

Implementation and system performance

To ensure continuity of production the AMACS system has been introduced in progressive steps. Being a modular build-up system, each of the modules has been developed as a stand-alone system and during introduction served as the fallback situation.

Phase 1 – Crane and shuttle-car computer system

This phase contained the installation, in several stages, of the crane and shuttle-car system for material receipt processing (Fig. 23). First, all mechanical transportation equipment and the buffer stillages were installed in such a way that material could be transported on roller tracks from the docks into the stores entry area under manual control. The next step was the completion of the shuttle-car trajectory and the installation of the automatic cranes. Hereafter the system was ready to operate in semi-automatic mode but still the JIT and stores entry buffer was not used.

In the last stage of this phase the PDP 11/23+ computer, handling the buffer administration, commenced operation. Then a full computer-controlled material flow was possible from receipt into the buffer and further via the repack/stores entry stations into the stores. Through keyboard control JIT material out of the buffer and other material out of the conventional stores could be directed to the exit tracks. All parts of the CSC system have been extensively tested for their functionality and finally a test has been carried out through which the overall capacity performance of this system was measured against specified criteria.

Some significant test results are given below:

- Data entry station in receiving area: 105 pallets/hour
- Automatic cranes (in/out): 124 pallets/hour
- Shuttle-cars: 178 pallets/hour
- Overall capacity performance: 97%

The downtime of the system elements as measured over a 19-week period is shown in Fig. 24.

Phase 2 – Line replenishment on pull base

Picking tickets for material issue are generated according to production programmes. The requirements planning was also the driver for issue of material to production (push). At first a 'Kanban' mechanism was set-up to enable material replenishment requests (pull) by production. The kanbans, identifying the requested part number, station number and quantity, could be put into mailboxes installed in the production

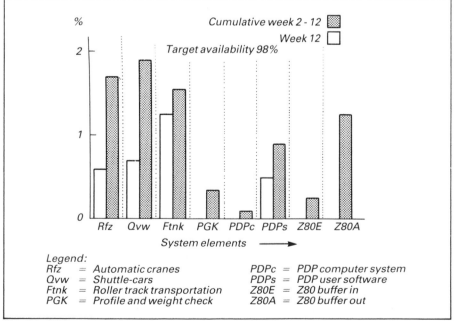

Fig. 24 CSC system downtime per element (1985)

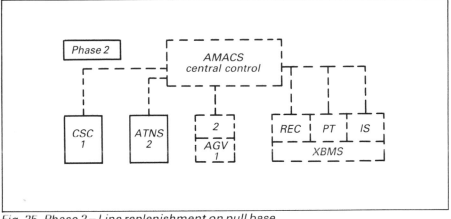

Fig. 25 Phase 2 – Line replenishment on pull base

area. Stores personnel collected the cards at regular intervals and took care of the material delivery.

Line by line the Kanban system was replaced by an electronic pull mechanism using the ATNS network in a stand-alone mode (Fig. 25). Menu cards with the appropriate request codes were given to the production operators.

Phase 3 – Optical AGV system

During Phase 2 of the project the floor wiring in the stores area and in approximately half of the production area (seven of a total of 13 aisles) was installed. After completion of the battery charge stations the semi-automatic delivery of tote-tins was started with three optical controlled carriers. Via a VDU keyboard located in the tote-tin store area, manual transport orders were entered in a microcomputer system thus enabling tote-tin rack transport from stores to approximately 100

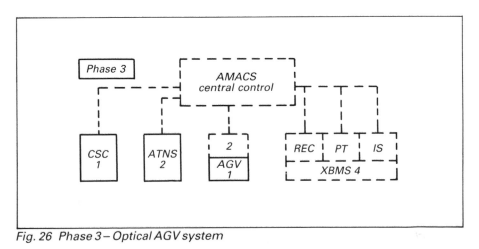

Fig. 26 Phase 3 – Optical AGV system

locations in the production area.

The communication between the controlling microcomputer and the AGVs was limited to one fixed order position and technically accomplished through optical sensors on the carriers. Once an order was transmitted, the carrier started to perform its mission without interaction with its environment and the controlling microcomputer (Fig. 26).

Phase 4 – AMACS – XBMS interface for receiving

Receipt transactions were entered through AMACS terminals. In this phase, AMACS performed screening on input and acted as a messenger system for data entry at the consolidators and the in-house receiving areas through to the live databases of the administrative XBMS system (Fig. 27).

Main achievements in this phase were:

- First usage of AMACS computer hardware.
- First interface operational between computer systems (AMACS/ Data General – XBMS/IBM).
- Data entry at consolidator's location.

To enable the partial AMACS system it was necessary to disconnect AMACS software modules from their nucleus and to build temporary program interfaces.

Phase 5 – Automatic pull cycle

In this phase, the three functional interfaces from AMACS to XBMS were installed:

- From Phase 4, the receiving module.
- The picking tickets processing (overnight spooling).
- Issue confirmation.

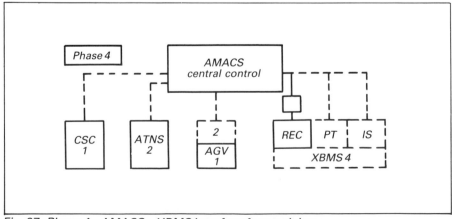

Fig. 27 Phase 4 – AMACS – XBMS interface for receiving

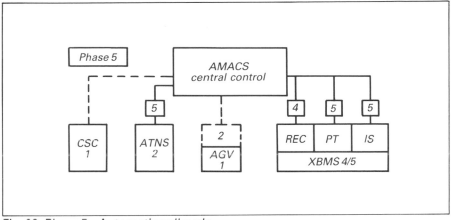

Fig. 28 Phase 5 – Automatic pull cycle

Transmission to and maintenance in the AMACS database of picking tickets, generated by XBMS, was started. The stand-alone pull system was interfaced to the AMACS software and automatic pull processing has been enabled as follows. The operator in the production station enters a part request code through his keyboard. After request validation AMACS allocates committed stock to that request by printing a picking ticket. Stores personnel pick and deliver material by AGV (tote-tins) or fork-lift truck (pallets) as in previous phases. Issue confirmation is entered in AMACS through manual terminal input (Fig. 28).

Phase 6 – Stores systems integration

Following completion of full AMACS-XBMS-ATNS communications, the next logical step was to link the physical stores transport system (CSC) to AMACS (Fig. 29). After extensive testing during the Easter

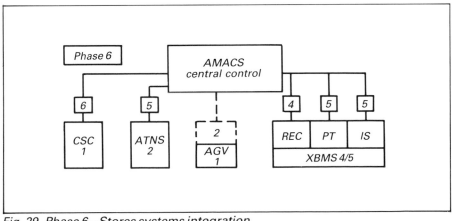

Fig. 29 Phase 6 – Stores systems integration

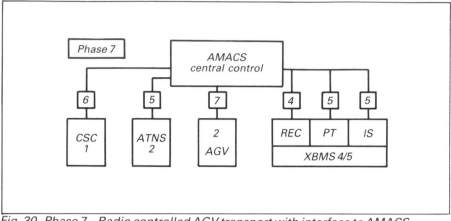

Fig. 30   Phase 7 – Radio controlled AGV transport with interface to AMACS

weekend (1984), the goods inwards buffer and roller track system were linked to the AMACS system. At receiving, this resulted in a simultaneous update of AMACS-CSC-XBMS data bases as receipt trancodes were entered through AMACS terminals. At the issue side, this new interface replaced the manual issue confirmation of individual picking tickets.

After implementation of this phase every pallet movement within the stores area was controlled by AMACS and its subordinate systems.

Phase 7 – Radio controlled AGV transport with interface to AMACS

During the previous phases the floor wiring was completed throughout the whole building and the PDP applications for radio controlled AGV transport were installed. Eight new carriers took over the transport from the three optical controlled carriers and transport orders were radio transmitted.

After interfacing with AMACS (Fig. 30), tote-tin rack and pallet transports were performed automatically upon arrival at stores' exit tracks. Pull request validation was extended to include reporting of

Table 1   AGV availability

		July	Aug	Sept	Oct	Nov	Dec
1984							
Plan	%	80	90	95	97	97	97
Actual	%	49	50	55	62	76	79
1985							
Plan	%	97	97	97	97	97	97
Actual	%	82	90	92	95	97	97

Note: In 1986, 13 carriers were operating at a satisfactory availability level.

successful completion of transport orders by the carriers.

Line-by-line box-pallet transport by AGV was actioned and after modification of the three optical carriers, all transport functions were performed by radio controlled carriers.

System availability measurements started in July 1984 and as shown in Table 1 the availability was only improving slowly.

Project time-frame

Starting mid-1982 with the development of a primary concept up and to and including a planned completion of the implementation of the AGV system (second phase) in February 1985, the whole project duration has been 32 months (Table 2).

Benefits achieved through JIT and automation

The automatic material control system has enabled considerable cost savings to be achieved. Reductions in material handling equipment and manpower are as follows:

- Fork-lift trucks: 52%
- Low-lift trucks: 100%
- Reach trucks: 28%
- Material handlers: 35%
- Administration personnel: 40%

Table 2 AMACS project milestones

		CSC	AGV Phase 1	AGV Phase 2	ATNS	AMACS
Supplier involvement/ quotations	start	11-82	11-82		06-83	04-83
Requirements specification	concept	12-82	12-82		06-83	01-83
	final	01-83	02-83		06-83	03-83
Functional specification	concept	03-83	02-83		08-83	05-83
	final	04-83	06-83		01-84	08-83
Ordering	start	03-83	03-83	06-83	08-83	05-83
	completion	06-83	03-83	06-83	08-83	08-83
Hardware & software delivery/installation	start	05-83	04-83	06-83	01-84	08-83
	completion	08-83	07-83	07-84	03-84	03-84
Functional tests	start	11-83	07-83	06-84	01-84	02-84
	completion	12-83	08-83	09-84	03-84	02-84
Capacity/Volume test	start	01-84	N/A	11-84	N/A	04-84
	completion	02-84	N/A	03-85	N/A	04-84
System implementation	start	08-83	09-83	07-84	03-84	04-84
	completion	11-83	12-83	02-85	11-84	09-84

Primary concept	June 1982–Sept 1982
Basic material flow data	Oct 1982
Operations analysis	Oct 1982–April 1983
Baseline system description	Dec 1982
Preparation of 'Capital Appropriation Request'	Dec 1982
Capital approval	March 1983

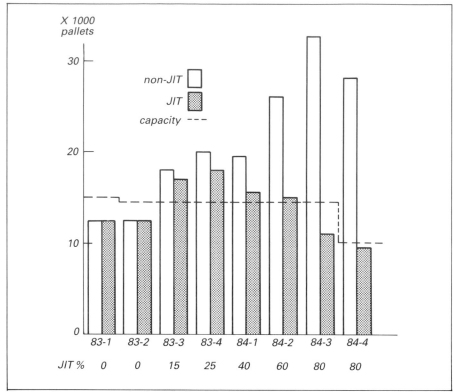

Fig. 31 Storage capacity impact by Just-in-time receipt

Space impact:

- Storage related space: 30%

Storage capacity impact of the Just-in-time strategy (Fig. 31):

- Pallet locations: 22–66% (by 40–80% JIT)

New United Motor Manufacturing

M. Sepehri
University of Southern California, USA

New United Motor Manufacturing Inc. (NUMMI) is the joint venture company established by General Motors Corp. and Toyota Motor Corp. to manufacture sub-compact automobiles in the USA.

The plant is in Fremont, California, on the site of an assembly facility which was closed by General Motors (Fig. 1). The total facility is 3,080,000ft^2, and includes a stamping plant of 180,000ft^2 newly built for the joint venture. The stamping plant which will supply hoods, doors, fenders and other major body parts has five stamping lines with 26 presses of 400–1400 tons.

The body and welding shop uses 170 robots to automatically make 90% of the 3800 welds in the body. The paint facility uses automated equipment and four robots, in addition to workers, to paint the vehicles (Fig. 2).

The final assembly line is approximately 1.3 miles, and operates at a rate of 60 vehicles per hour. From the beginning of operations in the body and welding shop through line-off in final assembly, there are approximately 1000 vehicles in work-in-process inventory.

NUMMI will produce 200,000 sub-compact automobiles annually for the Chevrolet division of GM. These vehicles, whose make is the Nova, are produced in four-door sedan and four-door hatchback versions. The product is sold only in the USA.

Acceleration plan

The company has been building cars since December 1984, and is going through a gradual acceleration plan. The current goal is to roll one car off the line every 54 seconds.

For the 1986 model year, the product is available in ten exterior colours (including two-tone paint treatment) and three interior trim colours, with a choice of two levels of trim. Approximately 15 different option choices are available, including air conditioning, automatic or manual transmission, and various types of radios.

Fig. 1   *Aerial view of NUMMI's plant in Fremont, California*

Design responsibility rests with Toyota, which controls all engineering information.

There are approximately 2000 part numbers released to NUMMI for manufacture of the Nova. The NUMMI production system is patterned after Toyota's production system, having been developed by Toyota as the first application of the famed Toyota Production System to automotive manufacturing in the USA.

Organisation

The plant is owned 50% by GM. In forming this joint venture, GM wanted to have a small car to sell in the USA and to find out how the Japanese production techniques, such as Just-in-time concepts, would work there.

Toyota, which owns the other 50%, was concerned about the quality and reliability of the supplier base in the USA; the transportation methods and systems, which are vastly different from the ones in Japan; and the mechanics of working with a labour force organised by the United Auto Workers (UAW).

The third 'partner' is the UAW. It was decided, as part of the agreement to form the joint venture, that the workforce would be

Fig. 2 Paint inspection area where each unit is thoroughly checked for any possible visible defects. After leaving inspection, the units are transferred to a 'selectivity' area to await scheduling for assembly. Approximately 1.5 gallons of exterior paint is used per unit

represented by the UAW. The UAW was interested in participating since it felt this type of worker-employer relationship might become common in the future, and wanted to get workers – most of whom had been on lay-off for over two years – back to work. Though most of the workers at NUMMI worked at the Fremont plant before and had been represented by UAW, the new organisation did not keep the old seniority system.

Since NUMMI is an independent company, its organisation includes administative functions (e.g. legal staff, financial planning) that are not normally located at an automobile assembly plant.

The board of directors consists of 50% GM and 50% Toyota representatives. Of the 63 management positions, 16 are filled by persons assigned by GM and 35 by persons on loan from Toyota, with the balance hired by NUMMI.

The Toyota people have the day-to-day overall management responsibilities for the organisation. Eight of the 12 positions for top management are held by people from Toyota.

The vice-president of manufacturing has responsibility for inspection, manufacturing and also process engineering. Product engineering is done in Japan by Toyota.

There are no logistics or traffic departments in the organisation. The function is part of production control, which reports directly to the chief operating officer and executive vice-president.

There is no total material management function, either. Purchasing is separate from production control.

At full two-shift production, the total workforce is approximately 2500 employees, of which 85–90% are hourly workers. The balance are salaried and non-union.

There are two classifications for hourly team members: Division I for regular production workers; Division II for general maintenance (skilled trades) and tool and die makers.

Corporate culture

NUMMI is a unique blend of Japanese and American culture. The team approach is reinforced through the use of open office areas, uniforms, first-come first-served parking, one cafeteria, warm-up exercises, in-plant sports facilities and team meeting rooms throughout the plant.

The workers are organised as teams. Each team has five to ten members, including a team leader who is an hourly worker. Three to five teams collectively report to a group leader who is a first-level, salaried supervisor.

The team leaders are not elected; they are selected through an interview process and are appointed by management.

Most teams get together on their own time before the shift, at break time or at lunch time. For training sessions, however, usually paid overtime is scheduled.

Most people wear NUMMI uniforms, which are optional. On a percentage basis, more of the salaried people in the office wear parts of the uniform than people working on the manufacturing floor.

The culture carries over into the office. The plant has open offices. The only really private office belongs to the president, who shares it with the executive vice-president. They also have desks at three or four other places in the factory.

There are many visual control tools in the plant. For example, training charts are posted in the group areas so that everybody knows the status of training. On a chart of the North American component area, magnetic stickers are used to move around the components and visually review the differences. Group attendance charts are also used.

Training

To further understand Toyota's methods specifically, and Japanese management techniques in general, all group leaders and team leaders (more than 300 people) were sent to Japan for three weeks' training at a Toyota assembly plant.

Toyota also sent about 200 persons from its plants in Japan to be trainers of NUMMI employees. These Toyota trainers spent from three weeks to four months training people throughout the plant and offices. Also, 24 of the management personnel assigned to NUMMI by Toyota, function as coordinators whose major responsibility is to assure the application of Toyota's production system through personnel training.

When NUMMI started there were about 5000 people available to be recalled at the Fremont plant, with capacity to hire only 2500 through a selection process. People were sent a letter describing some of the team themes and work concepts. About 3500 individuals replied. The applicants were asked to come in on their own time and spent about four days going through an assessment process.

The managers at different departments interviewed the hourly employees. After being hired, they started a full training programme.

Success concepts

Several concepts are emphasised in NUMMI's philosophy, and are central to the success of this manufacturing process. The concepts are similar to the ones practised at Toyota in Japan and are emphasised through orientation and training programmes and daily team meetings.

Jidoka

This is the automatic stopping of production equipment when an abnormal condition is sensed. Jidoka also refers to the ability of any worker to stop the assembly line when a defect or problem is found. Most machines have sensors and necessary infrastructure to identify a problem. When one occurs, a red light comes on and the machine stops. This concept also applies to non-automated equipment and labour.

When the line stops, production is lost. Therefore, workers are urged to identify the problem and take immediate action. They also have to develop countermeasures so that the problem does not recur.

Kaizen

This is continual improvement of machinery, material, labour utilisation and production methods through the application of suggestions and ideas of team members. Kaizen indicates there is no best, there is only better.

Baka-yoke

This concept refers to the use of devices or innovations which allow an operation to be 'foolproof'. The baka-yoke concept is used to make the critical process so straightforward that even a fool cannot miss it.

Muda

Waste should be eliminated from manufacturing. Waste includes those elements of production that do not add value to the product.

Muda, in particular, includes storing, inspection and moving parts. Therefore, the efforts are concentrated on inventory reduction, defect elimination and lead-time reduction.

Five why's

Instead of asking who, what, how, etc., the question 'why' is asked repeatedly to identify the real source of a problem. The concept is used to discover the root cause of a problem.

Heijunka

The production system is designed for balanced utilisation of all production resources, including workforce, equipment and material. Heijunka includes smoothing the production variety and volume over time by levelling fluctuations. It begins in forecasting and order processing, where variations are closely monitored and controlled.

100% quality

Total quality is essential to the successful execution of the NUMMI production system.

Using the principle of Jidoka and mechanical processes, where possible, quality checking is automated to identify conditions that are in conflict with specifications. Team members are their own quality inspectors, and are able to stop the line if a problem is encountered.

The concept of total quality also applies to material and components received from suppliers. NUMMI does not usually perform receiving inspection for quality or quantity on incoming shipments. However, shipments from those suppliers with identified quality problems are 100% inspected. The problems are analysed and corrected through joint efforts of NUMMI and the supplier.

The plant uses some inspectors at various production areas. Many charts are used to identify the specific items to be inspected. The charts are also helpful in training new workers.

Standardised work

Production is most efficient, quality is at its highest and waste can be eliminated when tasks are organised in a routine sequence.

The routine tasks of each team member are studied and a standardised work chart is developed. The charts visually depict the actions required to complete the tasks, including work operations, quality checks, walking, waiting and movement of material.

By measuring the time required for each action, the net time for total standardised work can be found for each team member. This net time is compared to take time (daily total operating time at 100%% efficiency/ required total production) to assist in workforce planning and efficiency improvements.

Standardised work charts are displayed at the job site of each team member. This allows team members to review their own jobs regularly. It also provides a tool for training new team members enhancing their cross training and Kaizen skills.

Kanban

The original meaning of kanban is signboard. Its current use has been expanded to include any signal which is used to pull material through a Just-in-time system.

At NUMMI, there are many types of kanban. In-process and signal kanban are mainly used in the stamping plant and in the body welding operations. Interprocess and supplier kanban are used throughout the manufacturing operations.

The kanban forms and formats vary according to particular needs. Metal triangle-shaped kanban are used in the stamping plant. Both re-usable and disposable paper kanban are used in component ordering and delivery. An electronic (computer to computer signal) kanban functions between NUMMI and its Japanese supplier.

Where applicable, the data on the kanban is barcoded to facilitate data collection. If the data change frequently, the information is handwritten and can be easily revised.

Small lot

NUMMI limits inventory through the use of small lot sizes in production, delivery quantities and lineside inventories. Frequent delivery is used both from suppliers to NUMMI and within the facility itself. Additionally, material handling techniques and equipment are designed and applied to facilitate the use of small lot size and frequent delivery.

Suppliers are encouraged to produce in small lots which conform to the daily or more than daily shipments that NUMMI requires of its suppliers. These small lots of material are further broken down when received at NUMMI, and parts are delivered hourly to lineside. The amount delivered is equal to the quantity necessary to support one hour's production.

There are six press lines, one large transfer press and two blanking lines. The presses range from 400 to 1400 tons. Two operators run each of the press lines and maintain the quality of parts.

In the stamping plant, and throughout the rest of the organisation, the emphasis is on quick dye change and rapid set-ups at the plant. The dye changes are put as close as possible to the machines to help minimise set-up times. The set-up for press lines is in the 6–8 minute range, which is twice as long as it is in Japan.

Japan sourced components

The majority of the parts for the Nova are sourced in Japan. The majority of these are manufactured by Toyota or its wholly or partially owned subsidiaries, the balance being manufactured by other Japanese firms.

Each month NUMMI provides Toyota with a forecast of the next four months' production requirements. These schedules are for planning, and do not represent a commitment for manufacture or purchase.

The requirement schedule for Japan sourced parts is based on actual parts usage at NUMMI.

NUMMI adjusts the quantities for miscellaneous requirements or known future events (e.g. scheduled overtime, holidays).

Packaging and transportation

Each day's shipment of components is gathered by Toyota and brought to the facility for packaging. The package is designed to provide optimum protection for the 10,000 mile trip from Japan to California, to maximise space and weight utilisation of the freight containers, to contain the quantity of parts necessary for hourly delivery to lineside and to fit the in-plant material handling equipment.

Parts are shipped in standard 40ft ocean containers which can be drayed on land when placed on a wheeled chassis.

The ships unload in Oakland, California, and the containers are drayed to NUMMI, a distance of 35 miles. Each container is identified with a control, or renban, number. The renban number is used to cross-reference the container and advanced shipment information listing of the container contents.

Material handling

At full production it will be necessary to unload more than two containers each hour. As the modules are unloaded from the containers, the individual boxes of parts are separated according to in-plant material delivery routes. These routes, developed through the use of the standardised work concept, are serviced with trains of wheeled carts pulled by tow motor vehicles.

The tow motor driver is responsible for unloading full cartons of parts at the appropriate lineside stock rack and picking up empty cartons. The delivery cycles are timed to occur hourly, with a maximum lineside inventory of two hours.

North American sourced components

NUMMI has 75 suppliers in North America supplying about a third of the parts for the Nova.

The concepts of the NUMMI production system are used to govern the supplier relationship.

Quality, price, location and other normal selection criteria are important. Suppliers must be willing to accept the constraints of a production system that may be unknown to them.

To assist the suppliers, NUMMI sends team members from production control, quality, manufacturing and purchasing in groups to visit suppliers.

Scheduling

Suppliers in North America receive a forecast of requirements every week. The forecast contains seven weeks of part number level shipping requirements. This preliminary schedule is for planning purposes only and does not constitute a commitment by NUMMI.

Two days before the shipment date, the supplier receives the final requirement schedule. This confirms the commitment to the supplier. Since almost all parts are shipped daily, the supplier receives daily final requirement schedules.

These final requirement schedules are based on the actual usage of the parts in NUMMI's production process. The usage is calculated by counting the kanban cards for material used in a day's production.

Packaging and transportation

The packaging for components supplied by North American suppliers was designed by Toyota to provide optimum protection during shipment, to maximise space and weight utilisation during transportation, to contain the quantity of parts compatible with hourly delivery to lineside, and to fit the in-plant material handling equipment. The majority of the packaging is disposable, corrugated cartons. Some returnable containers are used with some California suppliers.

A standard pallet configuration accommodates 11 different standard size cartons and allows a truck trailer to be 'cubed out'.

Only one part is received via railcar. All other parts are received via truck shipments.

Material handling

Depending on the location of the supplier, in-plant parts inventory levels may vary from two hours to two days. Parts from those suppliers shipping multiple times per day, such as the seat supplier, which delivers hourly, are delivered to lineside upon receipt. Other parts are received and put into storage.

The parts are pulled from storage through the use of kanban cards. The kanban cards are consolidated hourly and then collected by a material handling team member.

Tow motors and wheeled carts are used to move the materials to lineside, timed to allow hourly stocking of the line.

Upon completion of the delivery route, the driver drops off the empty cartons at the disposal site. He then returns the empty train of carts and prepares another load for delivery, based on kanban cards returned from production.

Concluding remarks

NUMMI is a test of the hypothesis that the Toyota production system can be successfully modified and implemented in a US automobile manufacturing plant that operates under a combination of Japanese and American management concepts.

NUMMI's team members throughout the organisation recognise that the experiment is not completed, and much more needs to be accomplished to achieve the goal of manufacturing the highest quality automobiles at the lowest possible cost.

Repco

F. *Bazeley and N. Baines*
Ingersoll Engineers, UK

Just-in-time (JIT) has in recent years attracted an increasing amount of attention in the Western world, where companies have become aware of the success of this manufacturing philosophy as practised in Japan. JIT is indeed a powerful tool in the hands of manufacturing industry anywhere in the world, but many manufacturers have come to regard it as being sophisticated and complex, and primarily suited to high-volume industries where suppliers can be more readily controlled and substantial economies of scale can operate.

Ingersoll Engineers holds a different view. First, it sees JIT as an essentially *simple* philosophy. Secondly, as equally applicable to manufacturing which is *not* high-volume and low-variety. In fact, this section of the engineering industry comprises 70% of the whole, and the majority of Ingersoll's clients are in the contract industry, or geared to small- and medium-batch production. This case study examines exactly what they can gain, why such manufacturers need to consider JIT carefully, and how it operates in practice. To illustrate Ingersoll's points, the experience and findings during a recent project undertaken with the Repco Clutch Company of Melbourne, Australia, are outlined.

Ingersoll's approach to JIT

Ingersoll's approach to JIT is out of the ordinary in that it emphasises the need to seek simple answers to manufacturing and business problems before considering investment in sophisticated techniques and complex manufacturing systems. The first stage in any project is always to consider how to make best use of existing facilities and technology by a critical appraisal of the current organisation. Numerous surveys have shown that it is the simpler changes and the lower technology applications which have brought the greatest benefits to the batch manufacturing industry to date.

That is not to say that simple answers are easy solutions. Indeed, the application of JIT demands the most detailed consideration of a company's present situation, coupled with careful analysis of the direction it will be taking in future. It may well be appropriate and cost effective to employ a flexible manufacturing system, advanced manufacturing technology or computer integrated manufacture – or elements of all three. However, these are only *tools* used to build up a fully integrated manufacturing system which operates in accordance with the principles of JIT. They are certainly not the starting point, which must be analysis and evaluation of need.

This thorough analysis must relate to more than just the company's manufacturing facility. Design, purchasing and manufacturing must all be seen as parts of an integrated whole, and Ingersoll believes it is a waste of time trying to apply JIT to any one part in isolation, for the delays between various stages of manufacture, from order receipt to despatch, are the cause of much lost profit. Another feature of JIT as advocated by Ingersoll is that it forces a basic reappraisal of decision making right back to first principles. For example, the product's design may be capable of standardisation or rationalisation, and it may be advantageous to change the form of raw materials used.

It is also necessary to consider the flow of information through the company. The principles of JIT must apply here too, since speedy response is essential, from order intake through to despatch. There is little point in the manufacturing department raising a materials schedule if the design is incomplete or purchasing is inadequately organised. If information is efficiently relayed, both through the factory and to suppliers, then not only is response time shortened, but, in addition, the necessity to carry large stocks is much reduced, and opportunities are provided for substantial cost reduction.

Application of JIT makes certain demands upon the manufacturer. First, he must ensure that the workforce is flexible and willing to respond rapidly to variations in the rate of demand. He must identify restrictive practices which are preventing the most effective utilisation of people, and he must take steps to stop these practices. People have to adapt from a highly structured, largely predictable environment to one in which aggressively rapid response to orders and quick decisions are the norm. It is not easy to change to such a radically new method of working, and here education is the key – *before*, not after, implementation of JIT, both to allow time for training before the new systems come into operation, and to increase the workforce's sense of participation. In this way, change is more readily accepted.

Other requirements may include accepting some under-utilisation of equipment and carrying a certain amount of 'safety' stock – depending on the nature of the business and customer demand. These concessions may be necessary to allow a degree of flexibility and to reduce lead-time, but the cost is minimal compared with the overall savings achieved by the JIT approach which Ingersoll advocates. With sensible planning, the key items of inventory equipment will be fully utilised, and levels of 'safety' stock will be greatly reduced.

Fig. 1 summarises how a JIT approach can have an impact on a company's business performance. It is particularly relevant in the small to medium high variety batch industry in that much of the benefit can be attained without massive capital investment. In fact, reduction of capital employed is one of the two key business advantages JIT offers. In all manufacturing companies, capital is tied up in inventory, plant and buildings. JIT substantially reduces inventories and thus releases space and capital for further, more profitable use. Materials, overheads and labour eat into profits, but a JIT approach can reduce these costs.

The second key advantage of a JIT approach is that the response to orders is much more rapid. Because the manufacturing facility is arranged to allow materials through the factory as quickly as possible, and because the information systems permit continuous, real-time assessment of stock requirements, allowing liaison with suppliers and scheduling to be completed rapidly, there is minimum delay between receipt of orders and shipping of the goods.

All these points may seem straightforward and simple common sense – but this is the essence of JIT. There are sophisticated techniques available to assist with the implementation of JIT – but the basic essentials come down to best business and manufacturing practice. In reality, how often are materials left lying around for hours, or even days, even though the actual process required may take only a few minutes? How often are high levels of stock held 'just in case' of supply problems? – but in fact causing other problems by taking up space, tying up capital and requiring manning to move; check and monitor them! How many costs are being *created* by existing systems? How much fat needs trimming to make companies 'lean and mean'?

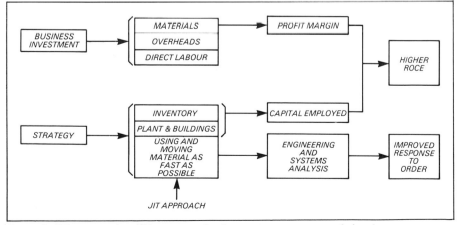

Fig. 1 A JIT approach will have a major impact on a company's business performance

Clearly JIT is an approach to manufacturing which is highly relevant to a wide cross-section of industry. This paper now turns from the general to the particular, and shows how the Repco Clutch Company (RCC) found JIT applicable to its needs.

The Repco Clutch Company

Ingersoll began working with RCC in 1984. It needed to assess its position and plan for long-term profitability in the light of changes in market conditions – in particular, the Australian Government's 'car plan'. The original car plan provided assistance to the industry over ten years, starting in 1965, by granting duty concessions to encourage an increase to 95% or more in the local content of all major vehicles produced in Australia. The components industry, including Repco, received a major boost and, over the ten years, the car industry achieved its greatest growth in both investment and output. Towards the end of the period, however, doubts about the economic rationale arose, and the success of the Japanese encouraged free trade exponents to work for a more international outlook for Australia's industry, in order to reduce the cost of vehicles to the consumer. Successive governments reduced their assistance to the components industry by a variety of measures, the most damaging of which to Repco was the introduction of export facilitation credits in 1982. This change in the car plan was designed to reward manufacturers who achieved increased exports by using duty-free imports of components to help them reduce the price of their cars. Repco was thus faced with a series of problems:

● There was a progressive reduction in the local content of Australian-made vehicles because of the export facilitation scheme.
● Intense international competition was deliberately encouraged by the export credits.
● Rising imports of spares and accessories, particularly from developing countries, were contracting sales and profits.

At the same time, massive changes in original equipment technology were taking place worldwide. Repco therefore had to make further investment to maintain quality at world standards, in order to remain competitive, while simultaneously achieving maximum economies of scale in order to remain profitable.

By the time the current car plan expires in 1992, it is expected that there will be three Australian vehicle builders rather than the current five, and model lines will be reduced from 13 to six. Rationalisation in the components industry is thus essential if Australian manufacturers are to compete with the world's largest producers with only modest tariff protection.

For Repco, rationalisation was indeed vital. It had developed as a company serving the interests of the whole Australian market in its product sector, with the result that it was eventually producing a tremendous variety of products. In this, Repco was by no means unique: like much of Australia's industry, its low-volume/ high-variety production was a direct result of serving a small but wealthy population and of the particular circumstances of Australian post-war development. It now had to answer some key questions about its future business strategy:

- Can our manufacturing units meet the terms of the car plan?
- Should we be manufacturing these products at all?
- How can we achieve the targets we need to remain competitive in the longer term?

It was to help assess the activities and investments required to answer these questions that Ingersoll Engineers was first consulted.

The business profile

Repco designs, manufactures and assembles clutch covers and clutch plates for a wide range of vehicles, both for original equipment and after-market supply. It also supplies components individually to the after-market. Orders from vehicle manufacturers for both original equipment and spares requirements are received directly into the factory. The factory supplies the after-market through the Repco distribution network, which is controlled through a warehouse facility. In 1984–5 sales amounted to A\$20.1 million and the company employed 207 people.

At the beginning of the project in 1984, the company's return on trading assets, at 28%, was good, but the implications of the car plan were that Repco had to improve further in order to be competitive worldwide. Ingersoll therefore recommended that the target should be set at 40–60%. Stockturn, at three per annum, was poor, and Ingersoll stated that a substantial improvement to eight stockturns a year should be the aim. Other areas of concern were long lead-times and high inventory levels. Tackling these problems successfully would reduce costs, and this in turn would improve the return on trading assets.

Ingersoll's initial observation of the business indicated that even the minimum overhead structure for this business needed an increase in manufacturing volume to provide the best cost reduction opportunity. RCC management saw 3% a year as the maximum volume increase feasible with current products. It was therefore logical to plan to introduce new products and Ingersoll, in conjunction with RCC management, calculated that this course should produce 5% growth a year, or 25% over five years.

Introducing new products would require both investment and space – but the rationalisation of products made necessary by the car plan and other market conditions provided the opportunity to produce the space within existing factory buildings, and to release capital by reducing inventories. The technology envisaged would have to be flexible and applicable to both current products and new products.

The way forward

A logical strategy thus began to emerge from the initial examination of the business profile:

- Introduce new products to achieve growth in the medium to long term.
- Rationalise the existing product range.
- Formulate manufacturing investment plans to modernise the facility, taking into account both the new products to be introduced and the need to reduce inventories.

A team was set up to undertake the task of identifying products which would be suited to the new equipment being planned for the modernisation of the factory. Ingersoll provided assistance by indicating the route RCC should take in this search.

Rationalising the product range was also closely related to the planning of the modernised production facilities. However, this planning had to take account of the fact that RCC, even after rationalising the product range, would still be in the high-variety, small-to-medium batch business. What was needed, therefore, was a facility capable of manufacturing the entire product range in small batches, with very short lead-times. For this, JIT was the logical approach to take.

The JIT approach

Product rationalisation and design

In all, RCC was manufacturing 720 varieties of clutch, 180 different cover assemblies and 110 types of diaphragm. Analysis of the business revealed that for clutch plates, cover assemblies and diaphragms, 80% of profit was coming from 5–12% of the products. For clutches and cover assemblies, 10–20% of the varieties were providing 60–70% of sales but the other high-variety low-volume products were using 50% of available capacity. If overheads were proportionally allocated, some of the low-volume products were being manufactured at a loss: in short, the higher volume products were carrying the high-variety low-volume ones.

Ingersoll and RCC together examined the variety of clutch plate components produced, and concluded that a reduction of 20% was quite feasible: materials used were low variety, and it was the high variety of components which was reducing machine utilisation and increasing lead-time. Opportunities for design rationalisation of cover assemblies were more limited, but Ingersoll recommended a reduction in pressure plate casting variety.

Reducing the numbers of these high-variety low-volume products had to be done with great care, in order to strike a balance between servicing the market and reducing the high cost of variety. This problem was tackled in a number of ways:

- Design rationalisation of some products was undertaken so that they used a reduced variety of components.
- A decision was taken not to manufacture certain products, as they could be supplied to customers more economically by importing.
- Where it was essential to continue supplying low-volume products, a detailed analysis was undertaken to see where margins could be increased.
- For some products, demand was so low that RCC could cease supplying them without serious risk of alienating customers or giving competitors an advantage.

Materials supply for JIT

RCC's suppliers can be broadly categorised as either foreign or local. Control over imports is necessarily constrained by travelling time, so that application of JIT to these supplies is naturally restricted, but the company has negotiated firm supply schedules from major importers, particularly of clutch facings from Japan and diaphragm steel from the UK. To reduce the variety of imported facings from Japan, and to increase flexibility of manufacture, RCC is considering the possibility of machining facings in-house and buying in only the raw material. This could have the added advantage of reducing capital tied up in inventory and allowing more flexibility in manufacture.

In the case of local suppliers, Ingersoll recommended setting up weekly schedules for parts arrival. Suppliers were required to operate to tighter schedules, but they would gain the advantage of increased volumes being ordered as a result of Repco's rationalisation. In some cases, it was found more cost effective to buy raw material bars to cut and process in the Repco plant than to purchase a variety of forgings in smaller quantities.

Operating to tighter, JIT, schedules will be assisted by the improved information flow coming as a result of the planned reorganisation of the information and control systems at RCC. The company fully appreciates that the success of a JIT facility requires good relationships with suppliers and it is now negotiating in order to agree firm schedules and delivery requirements, and to assess cost risks.

JIT manufacture

The objective here, as with other JIT activities, was to move materials faster through the production process. The main manufacturing operations could be broadly divided into press tool operations, machining and assembly. To reduce lead-time and operate JIT, the decision had to be taken to move from monthly to weekly batch production. Changeover time hence became a focus of concern, since with the production of smaller batches, an increase in set-up times would have the effect of increasing unit cost.

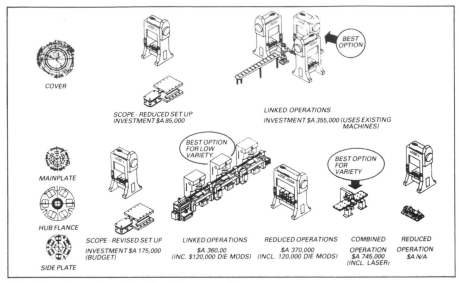

Fig. 2 Investment options for the press shop

With the objective of reducing changeover time in mind, workloads were calculated from the projected programme at each stage of the process, and improvements in manufacturing methods were evaluated. This resulted in a three-year investment plan of A\$3.2 million. One of the main aims of this investment would be to reduce set-up times by between 66% and 87% in press blanking and forming, hub and pressure plate machining, and in clutch and cover assembly.

To reduce changeover time effectively in the press-tool operations, it was possible to use quick tool–change equipment to simplify the task, and there was also opportunity to combine operations and hence reduce the number of discrete processes requiring changeover. Fig. 2 illustrates some of the manufacturing methods considered. In the case of machining operations, the combining of operations and automated handling was achieved by using CNC machines. For assembly, total automation was not justified, but each assembly operation was assessed for minimum changeover. For example, bench press tooling was redesigned to accommodate a range of products. In addition, relayout of the assembly area was designed to speed the provision of component parts.

Overall, the factory layout needed completely redesigning. The existing asembly flow was restricted, the toolroom was too far from the press shop, and stores took up 40% of the manufacturing space (Fig. 3). To streamline the flow of materials, products were put into 'cells' according to the 'family group' they belonged to, the effect being to reduce the manufacturing lead-time and the size of the components part store. Each cell will have its own machine setters for efficient changeovers, and relaxation areas have been included in the layout

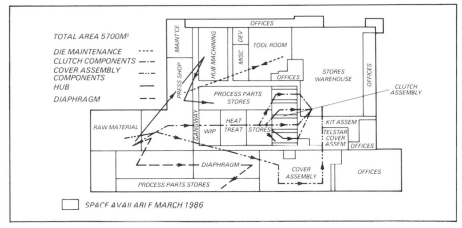

Fig. 3 A layout of the workflow shows that 40% of manufacturing space is occupied by stores area

close to the cells. These will also be used as areas where operatives will be encouraged to participate to improve the quality of both products and the environment. Relationships with the shop-floor are good, and flexibility presents no major problems.

The arrangement and location of stores was then considered, with the objective of integrating them fully in the work-flow pattern of the factory. For raw materials, it was decided that forgings and castings could be stored in pallet racking, but coil storage and handling would require special equipment. For processed parts and bought-out parts, narrow-aisle pallet racking was used for high-volume items with random location and man-rider access. Small parts storage was arranged for manual access. A kit marshalling area was designated for assembly. A despatch area was organised to hold buffer stocks of original equipment demanded by customers, and to group finished items awaiting transport.

Assessment of lead-times and of the opportunity to reduce them produced the figures and targets shown in Table 1. That such dramatic improvement is possible can be illustrated by current examples: for cover assemblies, the existing lead-time is 60 days for products requiring a total of 16 minutes' work in four stages; for clutch plates, six minutes' work in four stages also takes 60 days.

Table 1 Lead-time assessment

	Current lead-time (days)	Target lead-time (days)
Raw materials	23	17
Machining	3–5	1
Parts and process store	26	10
Assembly	3–5	1
Warehouse	27	10
Total	86	39

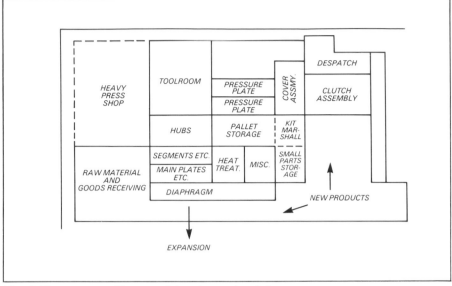

Fig. 4 The proposed layout of the planned activity

The need for JIT was evident, and the reorganisation of the layout was designed to help balance lead-time against delivery requirements. The greatest impact of this planned change in the layout was to release a large amount of space for the new products which RCC planned to introduce (Fig. 4).

JIT information and control

RCC's existing information and control system at the start of the project included a remote IBM mainframe computer with VDU access for enquiry. Stock control, works order control, purchase order control and costing, were maintained on computer. Information on material movement and order progress was provided each day, and manufacturing requirements planning was undertaken monthly. Capacity planning, sequencing and priority control were all manual tasks.

To provide JIT information leading to quicker order response, greater integration was required, and RCC decided to reassess its systems fundamentally. The computer system chosen would have to interface with the Repco distribution centre, and it would have to provide daily assembly schedules, weekly manufacturing schedules, a capacity planning facility, and the sequencing of weekly manufacturing batches to minimise changeover time. With this, the system could regularly update requirements data each week, or even more frequently if significant changes occurred between planning runs.

The information and control facilities provided would thus help the manufacturing facility to respond more flexibly and quickly to orders, to reduce changeover time by batching low-volume orders as far as possible, and to organise the sequencing of work in the most efficient way. Scheduling the flow of supplies would be better organised, and this would reduce the incidence of disruption. Physical parts control would also be tighter, and kitting would improve the efficiency of parts supply to the assembly line.

Two computer systems were considered for the job: a modification of the existing mainframe computer within Repco, and a new, stand-alone mini-computer which would be compatible with the distribution systems. The latter was initially more expensive, but the additional costs would be recovered within three years. In addition, this option offered several advantages over a remote mainframe, which included:

- User-friendly operating systems which could be handled by RCC personnel without data-processing expertise.
- Availability of complete, integrated business packages
- Up-to-date computer technology.
- An RCC dedicated system always giving priority to RCC problems.

The key objectives of the control system within the JIT environment could thus be summarised as:

- To increase the rate of information handling without taking on more staff.
- To allow 'real-time' updating and analysis of information.
- To integrate the sales, manufacturing and supply functions, allowing incoming orders to be rapidly incorporated into the manufacturing schedule.

SALES (CONSTANT 1984/85) A$ MILLION					
20.1 21.9	22.3	24.1	25.5	25.8	
1984/85 1985/86 HOLD AND PREPARE	1986/87	1987/88	1988/89	1989/90	
BASE	LEAN AND MEAN	ESTABLISH A JIT FACILITY	GO FOR THE MARKET	EXPAND SALES	EXPAND SALES
GENERATE CASH					
● VARIETY REDUCTION CLUTCH 58% TYPES COVERS 17% TYPES DIAPHRAGMS 43% TYPES ● RAISE AFTER-MARKET MARGINS BY 230K pa ● RAISE OE MARGINS BY 300K pa	● 8% OVERHEAD REDUCTION ● NEW PRODUCT SALES $1.5M	● 12% OVERHEAD REDUCTION ● NEW PRODUCT SALES $3.0M	● NEW PRODUCT SALES $4.0M	● NEW PRODUCT SALES $4.0M	
● RAISE DOMESTIC SALES 3% RAISE EXPORT SALES 3%	0.5% 3.0%	1.5% 3.0%	1.5% 3.0%	1.5% 3.0%	
● STOCKTURN 4 : 1	5 : 1	7.5 : 1	8 : 1	8 : 1	
● CAPITAL EXPENDITURE $2.1M	$1.0M	$1.1M	$600K	$600K	
● PROJECT EXPENSES $518K	$247K				
● IMPROVE LABOUR PROD. +4%	+ 7¹/₂%	+ 18%			

Fig. 5 The growth plan for Repco

SALES (CONSTANT 1984/85)						
A$ MILLION	20.1	21.9	22.3	24.1	25.5	25.8
	1984/85	1985/86	1986/87	1987/88	1988/89	1989/90
	BASE	HOLD AND PREPARE	ESTABLISH A JIT FACILITY	GO FOR THE MARKET	EXPAND SALES	EXPAND SALES
ROTC % INCL. PROJECT EXP.	19.0	20.4	34.5	44.1	53.7	53.6
PROFITABILITY A$ MILLION INCL. PROJECT EXP.	1.7	1.8	3.2	4.2	5.2	5.3
OUTPUT PER EMPLOYEE A$'000	84	88	98	110	113	114
INVENTORY A$ MILLION	5.1	4.3	3.4	2.5	2.5	2.5

Fig. 6 The financial model and the impact of the growth plan

Results

The result of this initial investigation was to develop and agree with Repco a growth plan (Figs. 5 and 6) using a financial model which set the business targets for the next five years and included a risk analysis.

The study began in February 1985, and the concept plan was presented in May. Detailed planning of the concepts was completed by September of the same year. In July 1985, Repco allocated internal resources for the future implementation of a JIT facility, forming an 'impact team' to develop and carry out the Ingersoll recommendations. A detailed three-year investment and timing plan was drawn up for implementing the JIT facility and achieving the 8:1 stockturn recommended.

An application in the automotive components industry

C. O'Brien, S. Chalk, S. Grey, A. C. White and N. C. Wormell
University of Nottingham, UK

This case study focuses on developments within one engineering company in the automotive components industry as it implemented major changes to its manufacturing organisation. An internal review of operational performance revealed significant non-value-adding activities, high levels of work in progress and static productivity despite capital investment. A new approach to manufacturing is described which has involved the integration of manufacturing facilities into a new purpose built factory, the adoption of a cellular approach to machine layout and the application of Just-in-time principles involving set-up reduction, total quality concepts and Kanban control.

The company is part of an international group and supplies original equipment components to the automotive industry. Before the change, it had two manufacturing sites some 30 miles apart vertically related in that one supplied steel pressings to the second which welds, paints and assembles the final product. It employs 600 people and is currently turning over in the region of £16 million per annum. It has a strong in-house design capability supported by CAD systems and prototype development workshops. As there is a trend for customers to subcontract design and development of new products, the company is in a good position to win new business.

Direct exports account for nearly 40% of company production, and this rises to 60% if vehicles assembled in the UK and exported are included. The company supplies most of the prestige car makers in Europe, and has supply contracts with major truck and commercial vehicle manufacturers.

The company operates very successfully in a highly competitive industry and has survived and grown by operating a niche marketing policy. Strong price competition has traditionally come from French and German companies, but a new threat is expected to emerge in the shape of Japanese supplier companies setting up efficient factories in the North East of England which will threaten the traditionally safe home market. Customer service levels are becoming increasingly important with requests for Just-in-time supply to the car plants. One hundred per cent quality assurance is becoming essential as customers grade their suppliers according to performance and move toward single sourcing of components.

During 1984/85, as part of a review of manufacturing performance outlined below, senior management of the company investigated similar manufacturing operations in Sweden and Japan. These visits confirmed that the company had successfully kept abreast of technological developments in manufacture but that there were alternative, and potentially superior, methods for organising manufacturing operations which should be explored. The setting up of a Teaching Company Programme between the company and the Department of Production Engineering and Production Management at the University of Nottingham provided an extra resource to help in the development of these alternative concepts and techniques into a practical manufacturing strategy.

Internal review of manufacturing effectiveness

During late 1984, the company conducted an internal review of its current manufacturing performance with a view to assessing its ability to maintain its present market position and to accommodate projected growth in the subsequent five years. It concluded that in the area of design it had a sufficient lead over its competitors, but that this was not the case for price, quality and levels of customer service.

The company had several areas of non-value-adding activity which should provide scope for significant reductions in manufacturing costs if directly attacked. These included materials handling activities, involving an excessive use of fork-lift trucks, and high levels of stock and work in progress. A survey of shop-floor space utilisation indicated that only 50% was allocated for manufacturing activities whilst the rest was used for gangways, storage space and quality control.

A safe and orderly environment was difficult to maintain. Manufacturing organisation and layout were based upon processes rather than products. Similar machines were grouped together in a functional layout (Fig. 1) and batches of components were routed through the different departments. Machines were capable of processing many of the types of components and it made economic sense to run large batches through because of the lengthy set-ups involved. Planning and control of production in this environment were complex tasks, and manufacturing throughput times were typically in the order of ten days. Quality of finished products was acceptable but this was only achieved after several layers of quality control had been applied. Levels of rework were unacceptable, as typically 20% of products were reworked at some stage in the production process. Operators were not specifically encouraged or rewarded for quality

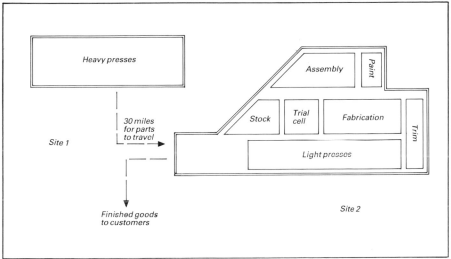

Fig. 1 Functional layout of manufacturing facilities at the old sites

work or 'right first time' activities. Preventive maintenance of machines and tooling was not adequate. It was not uncommon for press tools to be fitted to a machine, taking considerable time, only to be pulled out again after being found to be defective.

Customer service levels were good considering the company had a policy of holding no stock of finished product but this service was only maintained by expensively air freighting late orders and by excessive input of senior management time and effort. Lead times reflected the quantity of materials in the system and the volumes of work-in-progress hindered rapid feedback on process and quality problems, particularly between the two factories.

Some special purpose machines and robot welding had been introduced but projected benefits were often diluted in the overall manufacturing performance levels. Sales per employee had remained largely static over the previous five years despite the good levels of capital investment. Where management reporting systems were in effect they tended to concentrate on direct labour utilisation and an emphasis was put on controlling direct labour cost which accounted for only 13% of the manufacturing cost. The company retained an incentive payment system based on conventional time study techniques. This was tending to have a negative effect on the industrial relations climate within the company, and lead to the abuse of indirect booking systems and the proliferation of guaranteed payments for jobs with disputed standards. The management retained the incentive scheme as a means of controlling labour costs, although in reality this control was being eroded by relaxing standards on disputed times. In comparison, little attention was paid to controlling the 50% material costs or the 37% overhead costs.

Physical limitations on production capacity at both manufacturing sites meant that 30% of presswork was subcontracted out. It was felt that these constraints would impose limitations on new product introduction and company growth. The two existing sites did not lend themselves to expansion, were old, badly sited and not suited to presenting an image consistent with a quality producer of automotive products.

It was against this background that the company decided to take a fundamental look at its manufacturing strategy.

A new manufacturing strategy

A three-point manufacturing strategy was drawn up which over a period of five years starting mid-1985 would put the company in a position to satisfy projected customer requirements and would allow a doubling of existing turnover in real terms.

Firstly, the company would move away from its functional layout to create independent, and well-organised product cells run by teams of responsible, trained and motivated operators who had control of all the resources necessary for manufacturing a family of products from start to finish. The cell leader would act as a 'small business manager' and would have full responsibility for quantity, quality and timing of output within agreed schedules. The driving forces behind the consequent changes in job design and methods of work organisation were a movement towards group technology and Just-in-time concepts, rather than any ideal of job enlargement or enrichment, but an improvement in job satisfaction and the quality of work were seen from the outset as not merely a desirable outcome but a necessary prerequisite of making the required transformation successfully.

Secondly, each cell would utilise appropriate Just-in-time techniques and optimise material flows to achieve a five-hour manufacturing cycle. To support this environment, quick-change tooling techniques and planned preventive maintenance would be developed and total quality concepts would be strongly encouraged. Simplified planning and control systems coupled with small batch runs would facilitate flexibility and responsiveness to customer demands.

Thirdly, the whole production facility would be combined and relocated into a new purpose built factory in order to integrate all stages of manufacture. The new factory complex would be focused on the task of making quality automotive components at low cost and would be operative by mid-1987.

The success of the project would be judged on whether several targets could be met:

- A reduction of stock and WIP levels from a value of 10% of turnover presently, to 5% after two years and to 2.5% after five years.
- A reduction of rework levels from 20% of volume to 5% after three years.

- A doubling of sales turnover in real terms over five years. With a change in product mix and the introduction of new products this implied a 50% increase in material throughput with a factory 20% smaller in size.
- An improvement in sales per employee by 2.5 times over the five year period.
- Savings in manpower and overhead costs resulting from the synergistic benefits of integrating two remote factories onto the one site.

The detailed design stages necessary for the successful planning and implementation of the three-point manufacturing strategy were numerous and extended to all areas of the company's operation. For convenience of description, the main process redesign elements are split into three parts:

- General layout and cell design indicating how the product based layout is arranged in the factory and how cells were configured.
- Job redesign in the light of the changing roles and responsibilities of the operators and line managers.
- Changes in tooling and machine design requirements to facilitate cellular production techniques.

General layout and cell design
Allocation of products to cells

Having made a strategic decision to change over to manufacturing in product cells, the next step was to decide on how many cells and which products should be made in each cell. An analysis of the company's product range identified six generic design types having different process requirements. Marketing forecasts for existing and proposed products were obtained for the next five years and product vs quantity profiles drawn for each of the design types. It became clear that if cells were not to be too large to be manageable more than one cell would be required for some of the product families. Eleven cells were finally determined plus an area in which new cells could be set up for additional or replacement products. The number of discrete components manufactured in each cell and production volumes varied widely depending upon the market and the complexity of the final product.

After defining the cells in terms of the products to be made in them, it was necessary to calculate plant requirements for each cell. Using a modified process flow analysis in conjunction with estimates of future demand, bills of materials and operations routings, schedules of machine and tooling requirements were generated. Additional plant needed to be purchased to preserve the principle of self-contained cells. For example, an additional four 400-ton presses were required. To buy new equipment would have been prohibitively expensive and so the company adopted a policy of buying second hand equipment which they refurbished in-house.

General layout

The question of how these cells should be grouped together and how they should interface with the service areas was regarded as a determining factor in deciding the size and shape of the planned factory. Existing usage of area was carefully determined in order to have a sound basis on which to judge future area requirements.

Cell length and breadth were the next features to be determined. The process route of blanking → forming → welding → paint → assembly → packing gave an indication of length. Breadth was found simply by repeatedly designing layouts first with card templates and later using the company's CAD facilities to find the best arrangement in terms of material flow.

The role of the tool service facility was considered extremely important in the proposed method of operating. Press tooling was to be checked and passed off after every batch run and there would have to be a readily accessible tool bench and handling facility at the end of each cell.

The only process which could not be sensibly devolved into cells was the paint facility. The solution to this problem lay in the fact that all relevant products are painted at the same stage in manufacture, i.e. after pressing and welding but before assembly. The automated paint plant could therefore be positioned outside the main factory envelope and linked to each cell at the appropriate stage in the manufacturing sequence by a powered overhead carousel conveyor running the length of the factory.

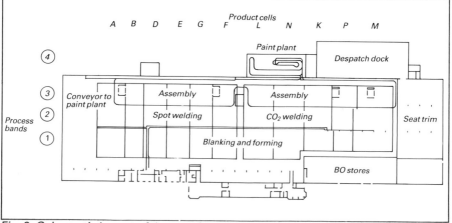

Fig. 2 Schematic layout of the new factory showing the arrangement of cells (A–M) and processes

The relative order of the cells was determined from several factors such as distance from goods in and out weighted by material volumes through each cell, projected cell life, etc.

A series of outline layouts was devised covering many configurations of cells, offices, goods inward, despatch, tool service and paint plant. Their relative merits were assessed using weighted factors, such as building costs, material flow, flexibility for expansion, etc. A diagram of the general layout finally adopted for the new factory is shown in Fig. 2.

Detailed cell design

Detailed cell design has centred on a policy of synchronising manufacturing output at all stages to customer requirements which are generally known several weeks in advance and relatively stable. Linearity of output at the various stages of manufacture within each cell is essential in creating a steady, uninterrupted flow of work and a balanced work load. Workcentres within the cell have to operate at a fixed planned work rate, in contrast to the traditional philosophy of running every operation at maximum output regardless of system throughput. With high operator flexibility within the cell, operators can move between workcentres as required, and where, to meet the required output, particular machine utilisation is low, it is possible for one operator to run two or three machines at one workcentre.

This idea of creating clusters of machines to match output was a fundamental feature of cell design and layout. The machines in the cluster have been configured in a 'C' or 'U' shape so that an operator can easily reach the various loading positions. Output flexibility has been enhanced by allowing enough space for a second operator to work in the cluster should the need arise. In many cases, rearranging machines has indicated how successive operations could be effectively combined by the use of dedicated manipulating devices or pneumatic cylinders with appropriate fixturing. Such opportunities had simply not been apparent in the previous layout. (An example of a cell layout is shown in Fig. 3.)

Operational control

One of the design rules adopted from the outset was that no fork-lift trucks would be allowed into the cells to handle materials. This was to save space and discourage a build up of work in progress. Transfer of components between processes has been achieved using small trolleys designed to hold half an hour's production. Daily production requirements are divided into multiples of half-hour batches and indicated on a board at the end of each cell. A simple Kanban system is used to 'pull' orders through the cell from the presses to final assembly.

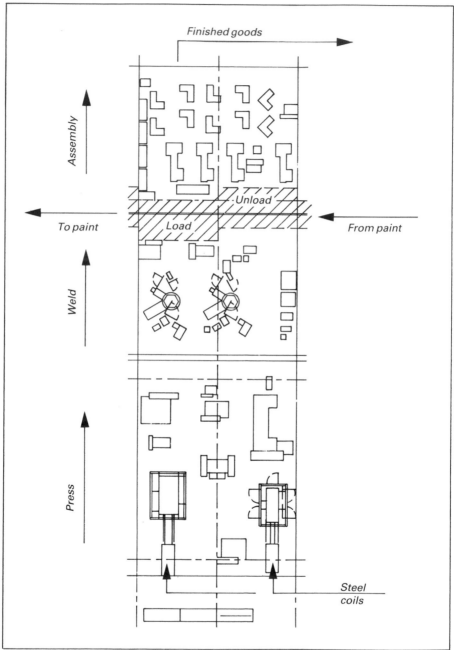

Fig. 3 Layout of a typical cell indicating the flow of work. A conveyor transports components to and from the paint plant prior to assembly

Strategic reserves of components are occasionally maintained alongside certain key 'bottleneck' machines in order that production flow can be maintained for an appropriate period in the event of a machine failure.

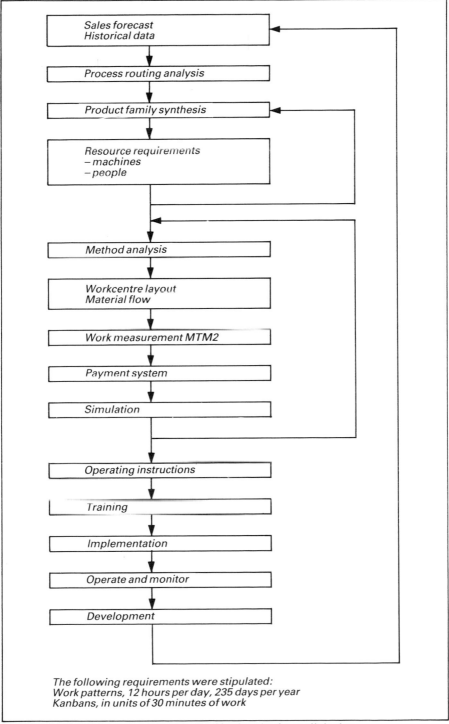

Fig. 4 The iterative procedure adopted in developing cell design

The policies to be adopted at each stage of the cell design were defined by a steering committee along the lines of the iterative procedure indicated in Fig. 4. Several pilot cells were configured in the old factory so that lessons could be learnt early on and in order that transfer to the new site could be accomplished with the minimum of modifications.

Job redesign for cell working

The traditional division of labour and work specialisation was no longer considered appropriate if the cells were to function efficiently. When working to a JIT system in cell manufacture, individual problems became group problems, as each operation is interdependent and a high degree of worker flexibility is essential. For this reason, it was felt that the situation lent itself readily to the setting up of a form of autonomous working groups, most well-known through their application in the Swedish car industry. In practice, however, it was not feasible to introduce, overnight, groups of operators with a high degree of autonomy due to the traditional working practices that had been ingrained in the workforce over the years, and which were reinforced through the payment system, traditional supervisory styles, a high degree of work specialisation and a hierarchical management structure. Specifically this meant that the workforce was not trained even to taking on the multiskilled role required of a cell operator, let alone the additional responsibilities of maintenance, tool service, production planning and group decision making. More crucially, management and supervision were not equipped to reinforce and facilitate such teams positively.

It was necessary then that the development of autonomous working groups was an iterative process so that employees at all levels were in a position to keep pace with such change. To this end a massive retraining exercise was initiated for all levels, covering not only additional hands-on skills, but also new approaches to work organisation and mechanisms for enabling and participating in group working.

A new system of shop-floor payment was devised, with the emphasis on high basic pay based on skills and knowledge acquired rather than work done, and with a group bonus for quality output. The basis of work measurement was changed from time study to MTM2, which was seen to be compatible with the shift in philosophy towards methods improvement, quality circles and the need for each operator not to maximise individual output but to produce just enough at the right quality, just-in-time.

A code of practice for cell working was devised giving a degree of autonomy which could be increased as the training proceeded. This was going a step further than the Swedish approach, having a much more specific definition of requirments. These formal rules were a rather paradoxical requirement for flexible group working but a very necessary one, as previous experiences with group working experiments within the company had demonstrated. What these experiences had also demonstrated was the need for clarity in the definition of the supervisory role and the various support services. With regard to the latter it was seen as particularly important to have members of particular support functions – production engineering, quality, tool service, maintenance, training – specifically accountable to one cell. Furthermore, the boundaries of each function and their relationship with the cell had to be clearly defined, and this often entailed the removal of demarcation lines and previous restrictive practices.

The following is a summary of the major changs in job design required on the shop-floor:

Before	*After*
Individual pay	Group pay
Output based	Quality based
45 operator grades	Two operator grades
Single jobs	Multiskilled
Traditional quality control	Quality audit function
Foreman	Team leader
Department (30–100 people)	Team (10–30 people)

Set-up reduction programmes

Quick-change tooling was viewed as essential in facilitating small batch runs and significantly increasing flexibility to meet fluctuations in customers' requirements. This involved tackling two main areas of manufacturing: special purpose assembly and fabricating machines, and blanking and forming presses.

Set-up reduction on a special purpose assembly machine

Very early on in the project the benefit of achieving fast set ups on special purpose assembly machines was assessed. As a starting point, a particularly important machine was identified as being a constant headache to all concerned with it. It typically took an hour to change it over from one product to another and involved the setter-operator adjusting mechanisms which were almost inaccessible. Because of the difficulties, people were keen to see the machine run for long runs between set-up changes.

A target of achieving set-ups in single minute figures was set. The first stage was to conduct a method study of the existing procedures, and a brainstorming session was conducted at which the setter-operator, the maintenance technician, the production engineer who designed the machine and a methods engineer analysed the existing method and put forward suggestions for its improvement. Some of these ideas, which were remarkably simple, were as follows:

- Replace bolted panels and doors with hinged versions.
- Eliminate nut-adjusted probes with quickly attachable interchangeable blocks of different lengths.
- Increase the use of dowelled blocks and wing-headed screws rather than bolts.
- Use of spring loaded pins and sliding blocks where clamping force was not important.
- Off-line cleaning of grease guns, using a back-up to keep the machine running.
- Off-line preparation of tools and kit prior to set-up.

The appropriate suggestions were adopted and implemented during an annual overhaul. The new method showed an improvement from 53 to 8 minutes. Costs were limited to proprietary steel components, some flame cut blanks and a few hours of technician time. The effect on the shop-floor was dramatic, changeovers were undertaken every few hours as sales required and the extra available capacity meant an overtime shift on Saturday mornings was eliminated.

Set-up reduction on power press tooling

Blanking and forming presses are an integral part of each of the product cells. If Just-in-time manufacture was to be achieved it became necessary to minimise set-up times to allow for short runs and frequent changeovers. This was in contrast to previous traditional presswork operations which encouraged long production runs with a minimum of tool changes. With 150 power presses in the company a standard approach had to be adopted, and to keep costs to a minimum, expensive solutions such as hydraulic clamping units had to be avoided. The exercise to reduce set-up times has been approached in three phases, each attacking a particular area of time-consuming activities.

The first phase has tackled the inefficiences of tool storage, location and handling facilities. These inefficiences were typified in the old production environment by tools often being stored over 50 metres away from the corresponding power press. This problem has been alleviated on the new site by locating all of a cell's tooling on specifically allocated storage racks at the press end of each cell. This allows tools to be transported by scissor jack to and from the presses in a fraction of the time it once took.

Phase two has involved identifying and separating those elements of tool set-up which can be carried out off the press from those elements of tool changeover which can only be carried out with the press stopped. It is these latter elements which have to be reduced if the time to change over from one product to another is to be minimised to allow the objectives of flexibility and small batch sizes to be met. On analysis, it was found that the traditional tool exchange operation relied upon a considerable degree of experience and skill. Often, on tool exchange,

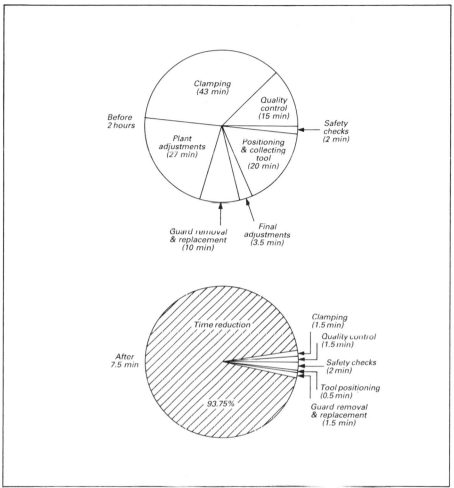

Fig. 5 *Typical reduction in tool changeover times which can be achieved on a 75 ton press*

press settings had to be adjusted, and with tools which are automatically fed, great care had to be taken in positioning the tools correctly in the press. The operation of securing tool die sets to the press was often effected with an odd assortment of strap clamps and t-bolts, which was both mechanically inefficient and time consuming. By providing suitably engineered systems of location on both tools and presses a very high proportion of the tool exchange time was eliminated. Clamping and location pegs are therefore being fitted at standard positions on presses to locate with corresponding lugs being welded onto the tools. In addition, tool heights are being standardised to eliminate all but final adjustments of the presses. By these means, the time-consuming elements of judgement and control in locating, fastening and adjusting tools at changeover have been much reduced.

The third phase of the exercise has attacked all aspects of the organisation of tool changeovers. These have included developing guidelines for the preparation of all hand tools, fasteners, materials required, etc. prior to stopping a power press for changeover, and by a redesign of the methods used in tool exchange.

The achievements of this three-phase approach have been considerable and although work on the power press conversions and methods changes are still ongoing, Fig. 5 gives an example of how tool changeover times can be reduced dramatically from two hours to 8 minutes on a 75-ton press.

Concluding remarks

This case study has sought to provide a snapshot of some of the issues which have arisen in one company's move towards cell manufacturing and Just-in-time production. The case is somewhat unusual in that it involves the reorganisation of a complete company. The changes were made at a time when the company was enjoying an excellent record of growth and profitability. It only came about because a successful management was prepared to look far enough ahead to where future competition might lie, and to review alternative approaches to manufacture which might better support the company's excellence in design and help maintain its competitive position into the second half of the decade and beyond. The company's objective is quite simply to match the efficiency of any producer in the world. It is the management's commitment to this end and its willingness from the outset to consult with and involve the workforce at all levels which have carried the project forward. Setting up the new factory and its systems of manufacturing has not been without its traumas, but the project has proceded to schedule and at the time of writing (May 1987) the new factory is fully operational and almost all production has been transferred into it.

It is still too soon to give quantitative evidence of how well the original targets for improvements in manufacturing performance are being realised. Some of the changes, such as the tool set-up reduction programmes, will take several months to complete. But already many benefits are apparent: work in progress has been drastically reduced; workflows are clearly much simpler and a greater sense of order prevails. As the workforce experience the new environment one senses that their enthusiasm for and commitment to the revised systems of working are growing. There is still a long way to go to complete the changes until the factory operates fully as intended. But the company has no doubt that the moves it has made are appropriate and timely, and that in making such changes it has secured a firmer foundation for a further successful chapter in its evolution.

Acknowledgements

The authors would like to acknowledge the directors of the company involved for permission to publish this case study, and the Teaching Company Directorate of the Science and Engineering Research Council and the Department of Trade and Industry of the UK for support of this project.

Hewlett-Packard

M. Sepehri
University of Southern California, USA
and
R. Walleigh
Hewlett-Packard, USA

Hewlett-Packard, a worldwide manufacturer of electronics devices and business computers, is known as a pioneer of Just-in-time and total quality control techniques in the USA. Several divisions of Hewlett-Packard have been very innovative in adapting these techniques to their operations.

The Computer Systems Division, located in northern California, has achieved impressive improvements in quality and inventory levels through a combination of Just-in-time and total quality control. The product, the Series 68 HP 3000 business computer, is a high-performance machine with low-volume production. It is the largest business computer Hewlett-Packard currently markets. A tremendous number of components go into the system, and the units go through extensive testing and a complex manufacturing process.

The Computer Systems Division is organised into the following departments:

- *Printed circuit board assembly* – Blank printed circuit boards (pcbs) and components are assembled by means of automated and manual component loading methods, wave and manual soldering, and various other operations, including masking, forming, back loading and assembly of fabricated parts.

- *pcb test* – Completed pcb assemblies are turned on and tested using both automated and manual test equipment. Defective and misloaded components, solder defects and internal board problems are identified and repaired in this area.

- *Cable assembly and test* – Cables and harnesses, which connect pcbs inside the computer, are assembled through the use of both manual and semi-automated techniques and equipment. Assemblies are also tested here and repaired if necessary.

- *Final assembly and test* – Subassemblies and fabricated parts are assembled here to make completed systems, which then undergo automated testing. Some defective subassemblies are repaired in this department; others are returned to manufacturing areas to be reworked.

- *Support departments* – Departments that provide support to the production areas include other processing, production control, purchasing, incoming inspection, stores, information systems, process engineering and production engineering.

Prior to June 1982, when the productivity improvement activities started, the Computer Systems Division was a typical manufacturer operation with a large number of back orders and inflated inventory stocks. A series of programmes were implemented, including statistical quality control (SQC) – which later became total quality control (TQC), machinery cycle time reduction using SQC, simplification of the process, a policy of not starting incomplete kits and a Just-in-time (JIT) production process.

When the total quality programme had been in place for more than a year, most process and quality problems had been eliminated and the manufacturing environment was ready for Just-in-time implementation. The production manager had read about Japanese manufacturing concepts and Just-in-time systems, and attempted to interest the rest of the group in implementing such concepts.

Initially, most people were sceptical about Just-in-time, and several argued against it. There were about 200 late back orders at any one time, and implementing such a system seemed impossible.

After attending a number of seminars about Just-in-time and visiting other plants that had implemented it, a few managers decided to investigate the applicability of the concepts in the plant. They sketched an overall work plan for reducing inventory and implementing a 'demand pull' system. Implementation was slow but steady, and the early results were promising.

Problem awareness

Before Just-in-time was implemented, the assembly process for printed circuit boards in the Computer Systems Division consisted of a number of steps, including gathering a kit of parts, autoinserting, hand loading, wave soldering, back loading and final testing. Boards were built in lot sizes from 20 to 200 and were controlled by work orders issued from the material requirements planning (MRP) system. The biggest problems were material storage and product backlog. Work was issued even if some of the parts were missing.

The MRP system, used for material scheduling, was based on batch quantities and average lead-times. The average production cycle time was three weeks, with actual time usually a few days longer. Production and materials groups were continuously expediting parts.

When the assembly process was started before all parts were in hand, the process would generally proceed as far as the wave-solder operation. If the missing parts had not arrived by this step, the partially completed boards were pulled off the line and stored on shelves. When the missing parts arrived, the partially completed assemblies were brought back to the line and the assembly process continued.

Data collected on parts back order problems showed that, on average, 98% of the parts were in the stores area when work orders were pulled. Materials management felt that because of vendor problems and variations in the production schedule, this performance was acceptable. However, from the point of view of the production department, it was the greatest obstacle to improving productivity.

Since most kits required as many as 100 different parts, numerous kits had back-ordered parts when they arrived on the production floor. Data showed that about 75% of all the kits pulled had one or more back-ordered parts when delivered and that, on average, each kit had one to three missing parts.

Production management had been working with the materials group for some time on improving parts availability, but had been unable to raise the in-stock level above 98%. Although management was aware of the back-order problem, it was not considered serious enough to make improvement a high-priority objective. Most similar assembly operations in the company were experiencing the same amount of difficulty in procuring parts. One reason for this was high demand in the chip market, which was causing a number of suppliers to miss promised shipping dates or to ration scarce parts to their customers.

After observing many problems with back orders and incomplete kits, the assembly manager and his group decided to implement a new process whereby no work order with missing parts would be started. They modified the parts-pulling process as follows:

- Stores should continue to pull kits of parts according to the MRP schedule.
- Complete kits were to be delivered to the assembly area as usual.
- Incomplete kits were to be placed on shelves in the hallway with notes indicating which parts were missing.
- When the back orders were filled, the completed kits would be delivered to the assembly floor.

The experiment was started on a Monday. Immediately, material began to build up on the shelves in the hallway. The work load in the assembly area began to slow noticeably. As work-in-process began to flow out of the area, supervisors showed some nervousness as their people became more and more idle. For the first week, very little new material flowed into the department.

Soon the material in the hallway became noticeable to higher division management. A compromise was reached under which the incomplete kits would be stored in a special area that could be more tightly controlled.

The work-in-process shelves gradually emptied as more of the old back orders were filled. The department manager decided that it might be possible to eliminate some of them. Significant pockets of vacant space opened up in various parts of the department, and it began to take on a cleaner look.

As the experiment went into the fourth week, the workers were still often idle even though work had begun to flow through the process again. A quick check of the production output showed that weekly production had climbed back to the level that was maintained before the experiment began. Incredibly, however, the amount of labour required to assemble a kit of boards had been cut nearly in half by this single process change.

As the department began to adjust to the new procedure, other problems began to surface. As work-in-process queues disappeared, an extreme variability in workload became visible. Sometimes the workers were almost idle; at other times they were inundated with work. Previously, the work-in-progress queues had hidden the variation.

Production control was called in to study the problem. As a result of the study, lot sizes were reduced significantly. Smaller lots of each board type would be delivered several times each week. High-volume assemblies would be delivered daily.

Additional attention was given to maintaining equipment. The reliability of critical equipment, and therefore its use, increased due to increased awareness of equipment mainteneance.

Data from before the experiment showed that if many boards were partially assembled up to the wave-solder step, when the missing parts arrived it would take approximately two days to expedite them through the process. New data showed that a lot of boards would be expedited through the entire process, from start to finish, in less than 12 hours.

The data showed a tight link between quality and productivity. Improved quality had eliminated the need for many complex processes. Less complexity meant that less work was required to produce a given output.

Total quality control

Total quality control (TQC) covers statistical quality control (SQC) and process enhancement activities. Key to TQC are providing feedback and understanding both the failures and the work that is completed. At the Computer Systems Division, information is returned immediately from the customer (i.e. the next station) on the types of failure, their locations on the board and their probable causes.

In pcb assembly, failure is traced to final assembly, autoinsertion or hand load stations. Failure information is entered into a Lotus program, and the parts per million (ppm) defect rate is plotted over time for different processes. Workers have access to microcomputers to retrieve information or to generate graphs to help them understand what conditions exist within a process.

The goal of the TQC programme is zero defects. In December 1984, wave-solder production reached the zero defect level for the first time. Although this was not repeated in the next six months, it proved that a zero defect level was achievable.

The Computer Systems Division began implementing SQC in its manufacturing area in July 1982. The purpose of SQC was to improve product and process and, in turn, increase customer satisfaction at minimum cost.

Thirteen months after implementation of SQC began, a number of significant improvements had been achieved, including a reduction in defects caused by defective solder joints, elimination of a large number of defects associated with the automatic insertion of components, and improvements in pcb assembly cycle time. For example, solder joint failure rate was reduced from 5000ppm to less than 10ppm a year later.

Automatic insertion-related problems were slashed from an astronomical 30,000ppm to a respectable 5600ppm. Similar dramatic improvements were achieved in cycle times.

The SQC process begins with the identification of a specific quality control problem. Data are then recorded and analysed using various charts and diagrams. Based on these data, process changes are implemented, and the results of these changes are analysed to determine their effectiveness.

The wave-solder project

The first project undertaken was an attempt to measure the quality of the joints being produced by the wave-solder process. Many defective solder joints were being detected in later stages of manufacturing, even though all soldered boards were inspected and touched up immediately after the wave-solder process.

The initial step was presenting a training class to the operators on the wave-solder team. They were instructed on how to collect data and plot points on X-bar and R control charts. The training emphasised that this was to be a team problem-solving effort and that management was firmly committed to the belief that most defects were caused by problems inherent in the process itself, and were not the fault of the operators. It was also emphasised that the operators' responsibility was to help identify problems and that corrective action would be the job of management.

After a few days it became clear that the process was severely out of control and that approximately 1.8% (18,000ppm) of the solder joints were defective. A process engineer was called in to study current operating procedures.

A matrix was developed to show the values of the critical variables, such as conveyor speed and solder temperature, that could be adjusted for each type of assembly. New operating procedures were then written and displayed. An immediate reduction in defects was noted as the new procedures were put into effect by the operators.

Pareto analysis (the 80/20 rule) of the defects showed that a large number were blow holes and that high defect rates seemed to be associated with certain lots of raw boards. A raw-board manufacturing problem was suspected, and a meeting was called with the supplier to discuss it.

The wave-solder group decided to begin incoming solderability testing of a sample board from each lot; if the sample showed poor solderability, the supplier was instructed not to ship the lot until the problem could be found and corrected. This procedure eliminated blow holes as a major category of defects. Two months of work and some simple process changes reduced the defect rate to approximately 2000ppm.

Further analysis of the data showed that insufficient solder in the holes was now the major cause of defects. Insufficient board temperature was the most likely cause, so an experiment was performed. After it was determined that modifying the equipment to increase preheat would be difficult, procedures were changed to run boards through the preheater twice. Although this practice was quite out of the ordinary, the fact that it was developed by the operators themselves helped ensure that it was followed faithfully.

Data on the revised process showed that other categories of defects seemed to have been reduced also. At the same time, the operators observed that the conveyor occasionally halted briefly and that boards that were on the conveyor during these halts tended to have more defects. Maintenance was called in and the problem was corrected. Control charts now indicated that the defect rate was lower than 100ppm.

Toward zero defects

As the wave-solder team became more expert at collecting and analysing defect data, a feeling was spreading that a defect-free solder process was possible. During the next few months a number of small projects were carried out. Solder rack dimensions were checked and a number of fixtures were either repaired or discarded.

It was determined that partially loaded assemblies which were sometimes placed on shelves for several weeks awaiting the arrival of back ordered parts often had higher levels of defects. A process change was made to reduce lead-time by delaying board loading until all

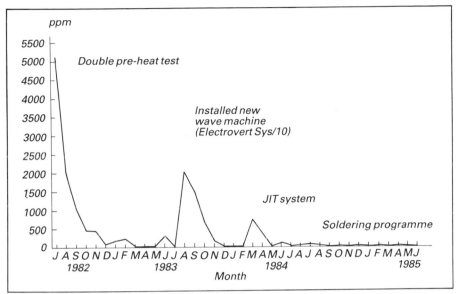

Fig. 1 In-process wave-solder defects

components were on hand. During the month of July 1983, the defect rate was calculated to be 6ppm.

With reduction of the defect rate and elimination of the touch-up operation, solder-related defects found in the turn-on area declined from 0.05 defects per board to 0.01 defects per board (Fig. 1). It should be noted that fewer defects were discovered in the test area than were counted by the solder operators. The criteria used by the latter were very tight, and many of the defects they counted were extremely minor.

Automatic insertion project

Another SQC project involved the process of automatic insertion (AI) of integrated circuits into boards. Kits of parts were delivered to the machine, and approximately 10–50 boards were loaded in a batch. After the components were loaded onto the boards, an inspection was performed.

When mistakes were found, the inspector replaced the incorrect parts. The machine often malfunctioned and caused parts to be misinserted or damaged. The machine operator spent a significant amount of time inserting parts and clearing jams from the machine.

The equipment was often out of service while the maintenance department worked to correct problems. Breakdowns lasting several days were not uncommon while replacement parts were air-shipped from the manufacturer's factory. Often large queues of work-in-process were formed, and weekend overtime was required to catch up with the workload. The process had been operating in this way for approximately two and a half years.

In September 1982, the supervisor of the AI department began introducing SQC techniques into the process in hope of improving the situation. A check sheet was developed so that the operators could begin collecting data. It listed 12 of the most common problems they had been experiencing. For this project only machine-related problems were studied.

The operators were given some basic instruction in SQC methods, and data collection procedures were established. It was found that defects were occurring at the rate of approximately 3% (30,000ppm) and that there was significant variability in the process. Pareto analysis showed that two classes of defect related to integrated circuit insertion accounted for the majority of the defects.

Based on these data, two changes were made in the process. First, maintenance was asked to add magazine cleaning and rotator adjustment to its regular twice-montly preventive maintenance procedures. Second, springs were checked, and worn or damaged springs were replaced. Further data collection indicated that these two classes of defects had been reduced.

Data collected for the month of November 1982 indicated that the major problem was 'phantom stoppage' of the machine. Poor design of the hole sensing unit in the machine was found to be the cause of that problem. An improved rotator had been designed by the factory and was ordered.

After several months of work, the mean defect rate was running at 1.5% (15,000ppm). Up-time on the equipment was now better than 90%, up from less than 50% at the beginning of the project.

While watching the insertion operation, the supervisor noticed that the boards tended to vibrate rapidly as the insertion head completed each insertion cycle. With assistance from the tooling department, support tabs were added to the tooling plate to improve board stability.

During this time, a concern was raised that the hole diameters on raw boards might be smaller than specified and could be causing misinsertions. A study of hole diameters, carried out by the operators, ruled out undersized holes as a defect cause. The rotator-to-preformer alignment was improved, and the rotation chamber was readjusted. These actions helped lower the defect rate to approximately 0.75% (7500ppm) for the month of May 1983.

In June it was noticed that the defect rate was higher than normal when plastic and ceramic parts were mixed in the same lot. Adjusting the interposer and the height of the preform chute allowed the machine to insert both types of part.

Turnaround time for batches of boards processed through the autoinsertion department was now less than five hours, down from an average of 20 hours earlier in the project. The defect rate recorded for June 1983 was 0.56% (5600ppm).

During the first nine months of the autoinsertion project, approximately 80% of misinsertion defects were eliminated from the process. The changes made to reduce defects produced equally dramatic improvements in turnaround and equipment up-time. The working environment was improved for both operators and maintenance people by management's emphasis on the collection and study of data, not on the blaming of operators and technicians for misinsertions.

SQC was also used by the production staff in the assembly area to find a way of reducing kitting errors. Before SQC was implemented, employees often spent much of their time filling out requisitions for missing parts, returning excess parts to stock and exchanging wrong parts before starting work on a batch of assemblies.

To reduce kitting errors, a team was formed of supervisors from both the assembly and stores areas. This group collected data using control sheets similar to those used in the AI project and then compiled a list of possible causes for the material discrepancies.

As a result of the team's activities, the material discrepancy rate dropped from 1.3 errors per kit to fewer than 0.2 errors per kit. In addition, assembly productivity improved, because less time was spent correcting material discrepancies.

SQC also improved the subcontracting process. The subcontractor had been allocated $180ft^2$ near the stores area for inspecting kits before they were sent out for assembly; however, as a result of improvements in the kitting process, the subcontractor was able to eliminate the inspection procedure.

Two important conclusions can be drawn from the results of these projects. First, significant process improvements can be realised in a short period of time through the use of simple, accurate methods of collecting and analysing data. Second, a team approach helps departments understand one another's working methods and make improvements that will benefit all.

Cycle-time reduction project

The first step in the cycle-time reduction project involved forming a team of managers and operators to analyse the collected data to determine the average cycle time of a single lot of boards. The standard lead-times used varied from 11 to 15 work days for each board type. For assemblies built by subcontractors, the average cycle time was 25 work days, with a range of 10 to 45 work days, as a result of the extra handling and transit time required.

A detailed process flow diagram was constructed to show each operation in the assembly and test process. Twice each day, a supervisor walked through the assembly area and recorded the location in the process of each board on the diagram. Because several lots of boards were in various stages of assembly at the same time, lot-number information was also recorded.

The data revealed that entire lots of boards often remained in the initial queue for several days before the first operation was started. When parts were on back order, no work was performed until the kits were complete, unless the work was late and needed urgently.

In addition, although most planning proceeded on the assumption that lots consisted of 30–70 boards, the production workers actually divided these lots into small quantities of 4–12 boards; these smaller quantities were then moved through the process at varying rates. In several cases, lots moved backward, which indicated some abnormal condition in the process.

A causal analysis of back orders revealed a problem in the procedures for testing random-access memory circuits: schedules for that area were not linked directly to the MRP system. In many cases, the incoming test department did not know when these parts were needed by the assembly area, which meant that deliveries were often late. After new scheduling procedures were created and explained to the buyers, the problem of late parts was eliminated.

To prevent component shortages, the purchasing department began highlighting potential back orders so that expediting could be started earlier. With the new procedures in effect, the average cycle time was reduced to five days, and work-in-process (WIP) inventory was significantly decreased.

The results of the attempts to reduce cycle times were summarised and shown to all supervisors and process engineers in the assembly department. This department was challenged to work toward reducing the average cycle time to less than 24 hours, which was considered feasible, because newly installed wave-solder and autoinsertion equipment had increased capacity sufficiently to reduce queue time in those areas.

Workstations were relocated to reduce material movement. Assembly teams were reorganised to allow continuous material flow through the process during both day and evening shifts.

As WIP queues were eliminated, a problem surfaced: work-load fluctuations were causing frantic activity during some periods and complete inactivity during others. After the production department studied the problem, it was decided to reduce lot sizes so that lots of each board type would be delivered twice weekly, or even daily in the case of high-volume assemblies.

As lot sizes were reduced to eliminate WIP queues, uneven material flow caused production areas to be lightly loaded for varying periods of time. During these periods, management reduced the emphasis on measuring labour variances and machine loading so that attention could be focused on levelling the work flow. For the most part, the productivity improvements brought about by reduced process complexity more than offset idle labour and machine time.

As a result of assembly process changes, the project reduced cycle time significantly. The average cycle time was 5½ days, down from almost 16½. In addition, sorting the data by lot size demonstrated that further improvements would be achieved as smaller lots reached the assembly area.

Key success factors

The following factors can speed the implementation of statistical quality control:

- Top management must feel the need to improve quality and must communicate this feeling to all levels of the organisation in order to establish a proper climate for SQC implementation. Gaining an understanding of the rapid progress being made by competitors, including the Japanese, in improving quality should help management internalise this need and feel a sense of urgency.

- Management must take an active role. Two kinds of support are required from management to implement a successful SQC programme. The first, which can be labelled 'passive support', is the support necessary to allow the programme to proceed: availability of competent statistical help; a sufficient budget to cover consulting services, training supplies and materials; 'blessing' the programme and letting the workforce know that it supports the programme and that rewards will be provided to those who are successful in achieving improvements.

  Active support must also be given generously. Managers should be as thoroughly trained in the principles of SQC as any staff member and should actively participate in projects as members of a team. Management must be prepared to take immediate action to solve problems when appropriate data are presented. It must provide an atmosphere that encourages workers to identify and report problems without fear of being criticised.

- Training must be ongoing and thorough. A long-term programme of training must be developed and the concepts practised diligently. Other professionals, including process engineers, should also attend the introductory training.

  As implementation proceeds, a long-term training programme should be begun so that the initial high level of motivation of the workforce can be maintained.

- Positive rewards for success must exist. It is even more important that participants are not penalised for their activities. Forming a team with representatives from a supplier who is thought to be supplying poor quality materials or making late deliveries can lead to substantial gains.

Current JIT environment

By implementing Just-in-time, approximately $2 million in just-in-case inventory was eliminated. Bringing active inventory to the production floor eliminated the need for raw material warehousing, and total space requirements were reduced by 46%. The materials move continually through the shop and are controlled at points designated as kanban control areas. These areas, located in various segments of the manufacturing flow, are basically material flow regulators. They facilitate use of a pull system that directs the movement of material to where it is needed when it is needed. As a result, only when a part is consumed or needed is the previous operation authorised to manufacture it. Data show that implementing kanban control areas reduced the amount of work-in-process inventory to $\frac{1}{15}$ of its size before implementation.

A set of large racks located behind the autoinsertion area are devoted to on-line material storage of integrated circuits (ICs). When operators see their workstation carts running low on materials, they 'go shopping' inside the warehouse or on the storage shelf and help themselves to the components needed.

The workers simply pull out tubes of ICs, and are not required to count the parts or track them. Parts are counted, verified and tracked when they are received. They automatically find their way through the assembly area and are post-deducted when the assembly and/or product leaves the area.

All inactive and some bulky parts are stored at an off-site warehouse. The majority of parts are small and do not take up much floor space. Bulky parts such as frames and power supplies are delivered daily from vendors.

Kanban, the tool for Just-in-time control, is simply a large rack with many slots. Each slot is assigned to a particular board assembly that goes on to the next level product. The boards arrive one at a time and are placed in the right slots. When the next operator needs the boards, they are picked up from the rack. When the board set has moved to the next area, the operators recognise the open slot and authorise the release of the next kit.

As a complete board set becomes available, the 'customer' (the next operation) is signalled by means of a green card. This shows from a distance that material is fully available. The customer may also signal a need for parts by showing a red card, and a complete kit of boards will be moved to the assembly area as soon as it is available. After the boards are taken to the customer area, the signal card is flipped back to indicate the start of work (Fig. 2).

There are also large signal boards which display the material status for another group that produces cable sets 60ft away across the building. If any of four dots on a board shows red, another set of cables is delivered (Fig. 3). After delivery, the red dot is changed to green.

Fig. 2 Red and green cards on the signal board in the bay assembly area indicate the status of each of the four boards used in the assembly

In the bay assembly area, a signal board indicates the status of each of four boards used in the assembly. If all dots are green, the previous operation should stop production.

Fig. 3 The use of signal boards like this one in the bay test area minimises the need for tracking of materials within the process areas

In the board test area, all boards are tested for functionality. Test stations are equipped to test all different boards one at a time with minimal set-up time. Therefore, the boards do not accumulate, and immediate feedback is provided.

The production build schedule is presented in a simple visible form on a board hanging in front of the production area. The master schedule is converted into a rate per day for each particular product. If production falls behind schedule due to a holiday or a machine breakdown, the extra work is scheduled for the following days so that the schedule can be met.

Actual versus scheduled time is shown on the schedule board for each day. A red dot indicates a demand that needs to be expedited. A green tab means the unit is already completed and has been moved through the area. This schedule is maintained by the operators themselves, who flip a tab to green once a unit is completed. No complicated tracking systems are required.

This board is a very common stopping point for many people. The production supervisors and managers know exactly what is done, and whether the planned schedule is being met. If there is a problem, the visibility encourages inquiry and hence provides for better communication throughout the organisation.

Parts and processes are highly standardised to minimise the type and number of set-ups. The computerised operational programs for automated equipment, stored on floppy disks, are easily loaded with a minimal set-up time of about 15 seconds.

Material is not tracked tightly within the process areas. Work-in-process is called raw-in-process (RIP) and is counted only when it arrives at or leaves a production area. However, any rejected components or scraps are returned to the system, since they are no longer usable for production.

Engineering changes are coordinated and handled carefully. Engineering groups such as test engineering and process engineering are involved in finalising the change and deciding when and where it should be implemented.

The maintenance organisation reports directly to production. It is absolutely essential to have immediate resources and attention available when a process breaks down. In a Just-in-time environment, a breakdown may halt the entire manufacturing operation in a short time. High emphasis is placed on preventive maintenance as opposed to fixing breakdowns.

Since the product line is not highly labour intensive, the labour portion of the production cost is minimal. Therefore, there is no need to keep track of labour extensively. Labour hours can be tracked periodically for financial statements, but in a Just-in-time environment, they need not be tracked against standards or used as a basis for performance compensation.

Cycle time is a key factor that is measured and monitored. The cycle time for board assembly at the Computer Systems Division was decreased to six and a half hours by August 1985, from more than 20 hours a year earlier. Cycle time includes the assembly time for 30 boards and 25 different part numbers as well as some queue time.

A key element of Just-in-time production is understanding and managing the process. Workers need to understand the process well enough to be able to improve it, document it, and take ownership and responsibility for it.

The characteristics of the process improve until it becomes consistent and predictable. Workers may still make adjustments or conduct further experiments such as adding or removing a production card. For example, for the wave-solder machine the workers measure and chart the heat and flux of board surfaces regularly to assure the soldering is consistently up to operating parameters.

Workers are allowed to try new things on an experimental basis and learn from the results. The charts are an excellent visible tool for exhibiting the results before and after a particular process is changed.

In July 1985, when the Computer Systems Division moved to a new building, it took the opportunity to enhance the layout and eliminate wasteful conditions. Travel distances between workcentres were further reduced, and material handling was improved.

The process was improved by combining two kanban control areas. This was done on an experimental basis, knowing that it could be changed back if it did not work. There formerly was a kanban control at the end of autoinsertion and one at the end of the wave-solder area. By eliminating the first one and simplifying the process, control requirements and the cycle time were reduced.

About 95% of IC components are autoinserted; the remaining ones are hand loaded on the board. The 5% hand loading reduces the level of complexity in the autoinsertion set-up.

Most parts move on a cellular type conveyor, which eliminates the need for additional movements or storage racks. With minimal adjustments in the solder machines, a variety of part numbers in lot sizes of one can be produced.

By working closely with the suppliers and implementing statistical quality control and Just-in-time at their plants, the purchasing department was able to cut the number of back orders at any particular time from 200 (with an average five-day late delivery) in 1982 to two (with one or two days late delivery) in 1985. Quality data and failure analysis have been fed back to vendors for the past three years, and now over 50% of components are shipped directly to the line. The responsibility for testing is left with the suppliers. Some vendors, particularly suppliers of large components, have local warehouses and deliver just-in-time.

Management considerations

The new environment is self-managing. Management does not have to be concerned with controlling the shop-floor, and instead can work on strategic issues, such as new products, new manufacturing strategies, newly developed markets, supplier selection and many long-term items. Production is an area of pride and not concern.

The move to a new building provided an excellent chance to clear up problems and fully implement the Just-in-time system. The old building was really a prototype, and the lessons learned in it were used to improve the layout design. A central warehouse was completely eliminated, and total square footage was reduced.

The factory was designed to be inexpensive to change. Only a few things are bolted to the floor, and there are very few walls. The production layout has a modular design, so that as new products are introduced or processes are improved, the layout can be changed quickly and inexpensively. The facility was moved within two days, and no scheduled delivery was missed. Three additional moves were made at almost no cost.

Just-in-time has brought many advantages. There is no paperwork on the floor. The number of defects per part assembled has gone down by an average factor of five per year. The problems and data are very visible now. Complexity has been reduced, and problems have been identified. The workforce is now a thinking workforce. Workers and supervisors identify significant opportunities for improvement and resolve problems. As a result, managers are relieved of having to deal with day-to-day problems and can devote more time to major long-term issues.

With the simplified control system – tracking the red and green tags – the production control department has experienced a reduced load. Its function is no longer expediting parts or assigning work, but rather process improvement, managing changes, product introductions and engineering solutions.

Lessons for implementation

Simplifying the process

Any ongoing business activity is a process with an output that can be measured and analysed. All of the activities in the process being studied must be documented. This is often very revealing, because the true process is rarely as straightforward as management assumes it is (compare Figs. 4 and 5). It is usually filled with contingency paths and rework loops that make it very complex. The objective is to identify process performance measures, collect data for them and analyse the data for consistency and sources of defects.

The more complex a process is, the more difficult it will be to achieve consistency in its operation and output. Total quality control seeks to

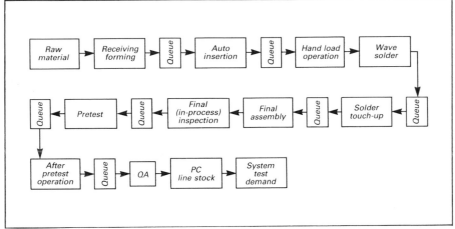

Fig. 4 Assumed batch material flow

find the sources of complexity so that they can be eliminated. Sources exposed by TQC analysis include defective raw material, late deliveries of components, erratically functioning equipment and poorly documented procedures.

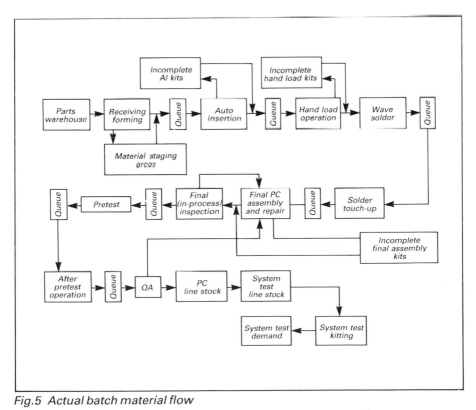

Fig.5 Actual batch material flow

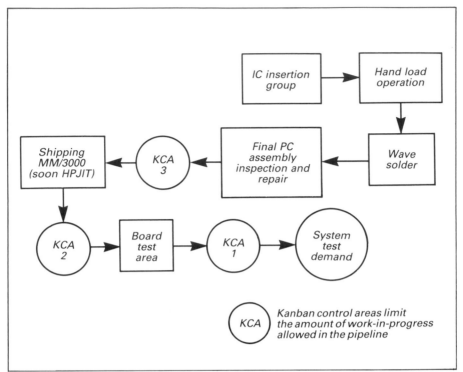

Fig. 6 Just-in-time material flow

Eliminating the sources of complexity eliminates the need for employees to develop creative and undocumented ways of working around them. The resulting simplified process will be much more consistent, reliable and able to produce a quality product.

JIT also greatly simplifies the movement of material through the factory (see Fig. 6). It eliminates the need for complex production scheduling of multiple levels of product subassemblies. Material is pulled as needed from the preceding production operation.

Problems with material availability become visible instantly. Communication between operations is better, because everyone understands what should be happening.

Eliminating waste

Any input of raw material, labour, capital or any other resource above the minimum required for the desired output is waste and will reduce productivity. The TQC methodology identifies the sources of defects and complexity in the process. Less material is wasted in scrap. Less labour is wasted because scrap does not have to be reworked.

JIT approaches the elimination of waste from another perspective. Because products are not assembled until they are needed, there is less inventory throughout the manufacturing process.

This has the direct benefit of increasing asset productivity, but the indirect benefits are even more valuable. With less inventory separating the various production operations, less time is wasted between the start of manufacturing for a product and its completion. The flexibility added by the shorter cycle times enables manufacturing to be more responsive to changing customer demands. A less obvious benefit is elimination of the wasted space which was used to store all the unnecessary in-process inventory.

Making problems visible

Both TQC and JIT promote the solution of problems by making them visible. After a process has been accurately documented, performance data are collected on the output of the process to determine how well it is operating. These data are analysed to determine the causes and seriousness of defects, and hard facts can replace opinions in the allocation of problem-solving resources.

JIT makes problems visible through a less direct, but often more compelling approach: stripping away the cover of inventory which is used to hide problems. Thus JIT forces managers to face problems and solve them.

Creating a climate for continuous improvement

Both TQC and JIT fundamentally alter the culture and operation of manufacturing. Their goals are zero defects and zero excess inventory, which can continually be approached. These goals must be accepted and internalised throughout the organisation, so that continuous improvement becomes part of the manufacturing routine.

Naturally, top management needs to be involved in and supportive of any major programme for it to be successful. On the other hand, the people who are closest to the work are the ones who know best what the problems are. These people are also often the ones who know best how to eliminate them and make operations simpler. Such changes will have a profound effect on their daily working lives. If they do not understand and participate in these changes, they may perceive them negatively and resist them.

Working on improving the process gives the individual employee a break from the routine of manual labour and also provides an opportunity to use thinking ability and knowledge. Most importantly, it provides a needed sense of ownership of the process they are involved in carrying out.

Kawasaki USA

R.W. Hall
Indiana University, USA

Founded in Tokyo in 1878 as a shipyard, Kawasaki Heavy Industries (KHI) for many years was best known for shipbuilding. In the century after its founding the company also became a leader in rail cars, locomotives, aircraft and aircraft engines, heavy industrial construction, construction equipment, steel structures, small four-cycle engines, robots and motorcycles. By 1980 it was a big, diverse manufacturing company with annual sales of about $2.3 billion.

At their Akashi Works in Japan, KHI's Motorcycle Division builds about 500,000 motorcycles per year of which 90% are exported. Like many exporting companies, KHI finds that it can improve its position in many foreign markets by building plants overseas. In 1975, KHI established a plant in Lincoln, Nebraska, to improve the product line and response offered to the North American market. As one of the first Japanese-owned factories in the USA, Kawasaki was determined as a matter of pride to make it a success.

In the USA, KHI is organised as Kawasaki Motors Corporation. At its headquarters in Santa Ana, California, are located research and development, marketing, sales and distribution. A distribution centre is also located with the plant at Lincoln.

In addition to motorcycles, the Lincoln plant produces snowmobiles and jet skis – products which have never been built in Akashi. Engines for the snowmobiles and jet skis come from Akashi, but over 50% of the other parts come from American suppliers. Both products require fabrication of miscellaneous parts and extensive body preparation prior to assembly.

Motorcycle production consisted of assembly with fabrication of frames, gas tanks and miscellaneous parts. Most engineering originated in Akashi, and new models started production in Akashi. One year after introduction, the tooling for models popular in America was shipped to Lincoln for follow-on production. The 1981 Lincoln production plan called for 75,000 units.

Kawasaki hired American managers to run the Lincoln plant by basicallyAmerican methods, but with goals and standards set by the Japanese and with assistance from Japanese advisors. The major performance goals were:

- Quality of production.
- Productivity (including the budget).
- Employee relations.

Things did not go well. Snowmobiles and jet skis started in small-volume job-shop operations which the Americans understood. However, the Japanese wanted to produce all products by the fast-flow repetitive production techniques used in Japan. More than language and culture prevented the Americans and Japanese from clearly understanding each other. They were simultaneously attempting two different approaches to managing the plant.

They had difficulty getting the right part to the right place at the right time. Inventory was randomly assigned to bin locations in the stores areas, and there was excess inventory. Parts might be on hand but could not be found.

Defect rates were high; rework rates were high. The quality of finished units was acceptable, but the cost of correcting the errors was too high.

Confusion, delays, expediting and rework led to poor morale. In early 1979, the United Auto Workers Union mounted a major drive to organise the Lincoln plant. Dennis Butt, the materials manager, became the third plant manager in the four-year-history of the plant, two weeks before the scheduled UAW election. After the customary dramatics, the workforce voted down the union 60–40. Attention turned to the plant's operating problems.

Gaining the workforce's confidence

Butt recognised that solution of the problems required management to work directly with the plant floor workforce. His first goal was to gain enough confidence from the workers to make this possible. That required a plan to end the turmoil in the plant. Under Butt's leadership the management adopted the following objectives:

- Get the right part to the right place at the right time.
- Put parts only where they are needed.
- Attain an uninterrupted flow of production, which requires:
  - reducing set-up times
  - reducing lot sizes and inventory levels
  - joining as many subassembly operations as possible to final assembly
  - doing it right the first time (improving quality)

The managers leading this programme had to work directly with the floor personnel. They also had to adopt the basic concepts of the Japanese Kanban system for getting the right part to the right place just in time for it to be used. That implies no inventory and no mistakes.

Some of the managers had always had proven results from 'hard-nosed methods', and they found it difficult to change. Almost all the managers were convinced that a materials requirements planning (MRP) system was the solution to their materials problems, and they found it difficult to change their materials philosophy. A reorganisation placed considerable responsibility on young managers in their thirties.

However, the problems at Lincoln were so serious that they needed a faster solution than MRP promised. Dennis Butt estimated that full development of the MRP system would have required a million dollars and an unknown number of years.

At the same time, the Kawasaki Akashi Works in Japan had just finished four years of Kanban development. The pilot project had been in their motorcycle plants, and results were so successful that Kawasaki Heavy Industries decided to implement Kanban in all plants suitable for it, including the Lincoln plant. A delegation from Akashi visited Lincoln to persuade the American management of the merits of Kanban.

Dennis Butt needed little persuading. He had spent two years personally rethinking the logic of optimal lot sizes. He knew that MRP implementation would be complex and time consuming, and that Lincoln needed a quick turnaround. The other managers had more difficulty accepting the merits of Kanban, but reluctantly abandoned the MRP project $300,000 into its development.

To reduce inventory, assembly lot sizes were set at 200, a lot size used for many items shipped from Akashi, and approximately one day's production on the main motorcycle line. The Akashi Works began shipping all large parts for each model in kits of 200 (or sub-multiples of 200) ready for movement directly to the assembly line at Lincoln. In that way it was also easier to keep all the parts for a model run together. Parts received from American suppliers were inspected, counted and repacked in 200s ready for the lines. (Parts from Akashi were not inspected. It was unnecessary.)

At the beginning of the model year a scrap allowance was set aside in stores for each model. When a part was scrapped in production, foremen could draw a replacement part from the scrap allowance. Then the lots of 200 parts stayed unbroken and as each model run finished assembly there were no parts left over. Work on the next model could begin very quickly with no time lost switching parts along the line. Fresh kits of parts were already positioned for the next model.

To get parts to the right place, floor locations were specified for each part. Pick lists for each model run were generated by computer from the line schedules. Material handlers used them to get the right parts to the designated locations in time for each model run.

This practice began on the main motorcycle line, and at first the operators and foremen were dubious. They anticipated that keeping the lots of parts intact would be bothersome. However, they soon found the extra trouble to be worth it because the practice reduced errors and expediting, the first major step in getting the right part to the right place at the right time. By May 1981 there were still occasional oversights, but the quality of life in the Lincoln plant had considerably improved.

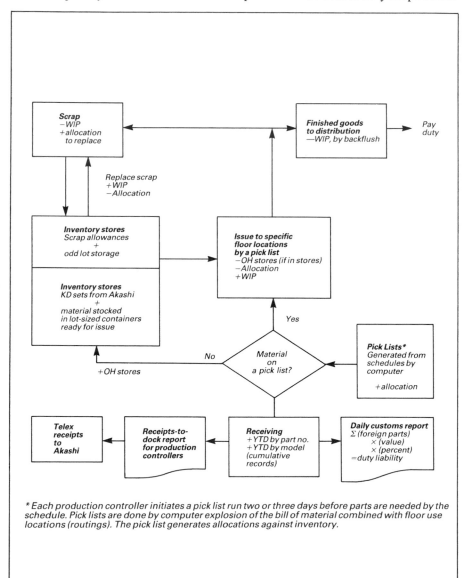

* Each production controller initiates a pick list run two or three days before parts are needed by the schedule. Pick lists are done by computer explosion of the bill of material combined with floor use locations (routings). The pick list generates allocations against inventory.

Fig. 1 Inventory control system

The assemblers quickly developed a liking for the 'Just-in-time' system, as they called this arrangement.

In the Lincoln inventory system shown in Fig. 1, the generation of pick lists is a key feature, but the Japanese advisors cannot understand why they are needed. In Akashi, material handlers are expected to know the parts needed without such a list, but at Lincoln there is difficulty getting the same person to pick the parts for the same area each day. With a low inventory, pick lists should be unnecessary, but as long as parts had to be found in a large storeroom, pick lists were useful. The American managers would use them again if they had to clear out a bulging inventory.

The Kawasaki Inventory System at Lincoln was very similar to a bulk issue system used by many American repetitive manufacturers. Issues from the storeroom decreased the on-hand balance and increased the work-in-process balance. However, the issues to the plant floor all went to a specific location, and the fixed lot sizes improved work-in-process accuracy. Sometimes parts went directly from the receiving dock to the line, and Kawasaki wanted to do that with all parts.

The Lincoln plant had almost 30 days' inventory on hand in May 1981, much more than wanted. Only about five days' inventory was actively used in production. Most of the rest were parts left over from previous snowmobile production.

Cycle counting had been initiated, but not fully developed. The official inventory was taken by an annual count. Cycle counters checked samples of items selected at random from inventory. If the on-hand balance was within $\pm 0.5\%$, the record was considered accurate. On that basis, motorcycle parts were considered to be 90–95% accurate and snowmobile parts about 70% accurate.

The acuracy of the inventory records was very important to Kawasaki in June 1981. They were audited by US customs officials to qualify as a permanent Foreign Trade Zone, which meant that duty on parts imported from Japan and elsewhere did not have to be paid until the finished motorcycles left the Lincoln distribution centre. Customs officials had to be certain that all parts entering the Lincoln plant would be fully accounted for as they left on assembled motorcycles. The bills of material had to show the country of origin of all parts and accurately reflect all engineering changes. The inventory records of foreign-sourced parts had to be very accurate. The Lincoln plant passed the audit with flying colours – the first manufacturing plant in the USA to do so.

Much of the improvement in inventory accuracy came from the issue of standard lots to designated floor locations. The odd lots with non-standard counts were kept in stores, and the decreased level of inventory helped keep straight what was on hand. Using space released from stores, the Lincoln plant could expand production without extending its walls.

Reducing set-up times

To run the smaller lot sizes, the Lincoln plant drastically reduced set-up times in key operations. Ideas came from Akashi and the Japanese advisors, but the Americans liked working it out for themselves. Dennis Butt was authorised to spend money on general projects to reduce set-up times. There were no case-by-case studies or formal approvals. With the key objectives in mind, what needed doing was obvious.

Quick changeovers began on the final assembly lines. It began by issuing the standard lot sizes of 200. Then physical alterations enabled a quick transfer from one model to another by rapid tool changes. A few gravity feed racks were introduced. These sequenced parts for model changes to slide forward to the assemblers as they were used. The model changeover time was reduced from hours to five minutes.(However, on the main line on the day of the author's visit a changeover on the main line was delayed 1½ hours to locate a part. The changeover procedure is still being debugged.)

In order to prevent building up inventory in fabrication areas, projects to reduce set-up times began in those areas. Almost all changes to equipment were performed by plant personnel. For example, the jigs required to weld the frames for various models were left set up in different welding booths so that welders transferred from one booth to another to change models. The welders themselves did most of the work setting up the booths in space created by the reduction of inventory.

The quick set-up method which attracts the most attention from visitors is a carousel track for dies used on a 200 ton Bliss press which bends frame parts. This arrangement is sketched in Fig. 2. The dies used for pressing different bends in the frame tubes are located on the roller track circling the press in the same order in which they are required. The press operator does all the set-ups by himself. Each die change takes two to ten minutes depending on the die.

Very important in this die change is the use of pins and shims to quickly align the die halves in the exact lateral and vertical positions. With minimum physical effort the operator can position the die so that the first part made is a good one. That saves both adjustment time and tubing scrap, a necessity to make frequent die changes practical. The operator checks the first piece bent after each set up, and an adjustment is rarely necessary.

A number of changes in the fabrication of gas tanks reduced set up times and made it more a line flow operation. One 500 ton press had the die change time reduced to under one hour by improving the methods of die handling and alignment. Under development is a table drilled full of holes on its surface with an air chamber underneath. The resulting 'air hockey effect' allows the die to slide in and out of position with minimum effort.

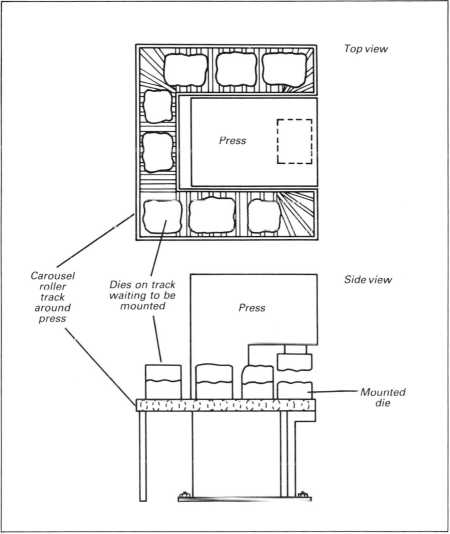

Top view

Press

Side view

Carousel
roller
track
around
press

Dies on track
waiting to be
mounted

Press

Mounted
die

Fig. 2 Quick change of dies on bliss press sketch of carousel method

The most difficult step for American engineers was to think of simple changes. The idea of the Japanese was to spend very little money and make the changes using plant personnel. However, organisational instinct is that a project is more important if there is a formal study and several thousand dollars involved.

Having workers perform multiple functions is also very important in making quick changeovers. The idea of quick set-ups is to simplify the process so that operators do not have to wait on mechanics to make changes. In the Lincoln Plant the maintenance and tooling personnel kept busy modifying equipment to simplify changeovers. The level of true responsibility of each job was therefore upgraded, and by early

1981 the Lincoln workers were beginning to pull together to achieve substantial improvements.

Balancing operations for straight flow production

To reduce the buffer stocks between operations, the rates of production of all operations needed to be balanced. By May 1981 several subassembly operations had been tied directly to the final assembly lines, saving space and inventory. However, the fabrication rates of frames and gas tanks proved more difficult to equate with final assembly rates. The welding rates of KLT frames were matched with assembly, but the frames for the larger models were not.

The gas tank was the most cosmetic piece on each motorcycle and the reject rate on tank preparation was high. In addition, a large percentage of spare parts work had to be mixed with production of gas tanks for assembly, and there was difficulty with the adhesion of paint to the tanks. The quality of the pre-paint wash water at Lincoln was such that a paint consultant was called to advise the type of paint to use. A type of paint different from that used at Akashi was selected, and the reject rate dropped below 20%.

Another improvement was to establish better coordination with the decal companies supplying the trim and nameplates for the gas tanks. This consisted of educating them to send only the number of decals required to arrive on the date designated with no over-shipments or under-shipments. Lincoln used its own scrap allowances in ordering these. The decals were expensive and had a short style life. Part of the problem had been in making substitutions, trying to work down inventories of obsolete decals, or substituting for new style decals that had not arrived. Each of these practices contributed to expediting as a way of life in tank preparation and to a high reject rate. Eliminating them reduced the unpredictability of tank preparation and brought closer the possibility of balancing tank preparation with the assembly line.

Solving these problems took time. Sorting of painted parts was reduced by building paint hangers which would hold all the parts for a given model which needed the same colour paint. Since they were kept together as a group through the paint drying conveyor, it was easy to keep them as a group in sending them to assembly. It also provided a check that all the parts for a model were in preparation and going through paint when the model was set up on the assembly line, another way of getting the right part to the right place.

However, problems remain. Weld rates on all frames do not match assembly rates. It takes extra inventory to buffer the problems of tank preparation from assembly. The Lincoln plant had hoped to have some kanban cards in circulation between final assembly and these areas by June 1981, but the problems proved more difficult than supposed. A true 'pull system' cannot be implemented without solving them.

Quality improvement

A programme to improve quality is one of the major ways to reduce the interruptions in production. This began on the final assembly lines with the directive to stop building 'cripples'. A cripple is a unit that requires added work or rework after it leaves the final assembly line. Cripples are caused by parts shortages, improper assembly or bad parts. In any case the results are about the same. Mechanics off line must make the corrections, often by tearing down some of what has just been assembled. This is obviously costly, but it is a common practice in many assembly operations.

The changes at Lincoln were intended to 'do it right the first time'. The 'rover' mechanics assigned to each line went wherever an operator was having trouble. Foremen or group leaders assisted if there was a problem. If assembly became difficult, line speed was reduced. If it resumed smoothly, line speed was increased. If all this did not prevent improper asembly, the line was stopped until the condition was corrected.

Assembly operators could not do their jobs correctly if they never knew what they were doing wrong. Inspectors or others who found defects – faults in the units for whatever reason – determined the cause of the problem and gave immediate feedback to the operators on what the problem was. The longest this feedback took was one day. Even if the defects were found at final inspection, the cause was on a report for the assemblers the following morning.

The immediate feedback was most important in determining how to correct problems, and it took only a short time for operators to realise that keeping up with the line was less important than installing a part correctly.

The total number of inspectors required decreased. Quality control inspectors no longer worked on the assembly lines. Four inspectors performed final inspections and eight more worked receiving inspection. (No parts received from Akashi were inspected. Parts from American suppliers required inspection.)

The scrap reporting system encourages the open reporting of defects. Scrap from all sources is reported on a simple form and placed in a gondola on the plant floor. No foreman is 'charged' with scrap, so there is no reason to hide it or to delay reporting it. Accurate scrap reporting is emphasised for three reasons:

- The causes of scrap must be analysed.
- It is necessary to get credit from the foreign trade zone.
- The pick lists and work-in-progress inventory must be updated.

The most important change was getting people to accept responsibility for their own work. The results of the changes made on the KLT line in early 1981 can be seen in Fig. 3. Defects from jet ski assembly

January 22, 1981 KLT Final Inspection Summary		May 11, 1981 KLT Final Inspection Summary	
Number of units inspected	82	Number of units inspected	63
Number of units without defects	16	Number of units with defects	55
Number of defects	116	Number of defects	9
Defects found at first inspection	67 (58%)	Defects found at first inspection	4 (44%)
Defects found at repair	37 (32%)	Defects found at second inspection	5 (56%)
Defects found at final inspection	12 (10%)		
		Average number of defects per unit	0.14
Defects per unit	1.42	Average defects per unit at first	0.06
Defects per unit at assembly	0.82	Average defects per unit at second	0.08
Defects per unit at repair	0.45		
Defects per unit at final inspection	0.15		

Fig. 3 Comparison of inspection reports before and after installation of 'zero defects' quality control on the KLT assembly line

plummeted within one week after the line foreman understood that it was the responsibility of his crew and himself to present a finished unit that could be handed to the customer on the spot.

A number of employee suggestions improved both quality of product and straight-through production flow. One of the best was an idea to perform the 'run up' test of the KLT motorcycle as the last station on the assembly line. That avoided the running test of the finished motorcycle after assembly. It also provided very fast feedback to the line workers on problems observed. By May 1981, the number of KLT motorcycles which needed rework after leaving the assembly line was down to about five per day. The Lincoln plant began to think about reducing this number to zero so that KLT motorcycles could be created at the end of the assembly line. (That is not done at Akashi, nor is it being considered there.)

The quality improvement programme had the characteristics of the original zero defects programme (though not the 'hoopla' which later became associated with it). There were no quality control circles, nor did anyone have plans for them, and consequently the Lincoln management considered their quality programme as basic good management, not as a body of techniques borrowed from the Japanese.

The KLT three-wheel motorcycle and Japanese design methods

The KLT is a short-wheel base, low-speed motorcycle for all-terrain use. It is easy to ride and rugged enough to use on rough ground. It is promoted to ranchers and outdoorsmen as something of a replacement for a horse. Sales are concentrated in the Western USA.

Sales forecast
- By models in monthly buckets
- One-year horizon
- Three-month replanning cycle

Sales plan
- A set of requirements by model on the finished goods warehouse at Lincoln, which is fed by the Lincoln Plant. To get this, the forecast is 'adjusted' by Akashi
- One-year horizon
- Three-month replanning cycle
- Closest three months are firm
- The requirements of the Lincoln Plant are based on this
- Plan is approved by the plant manager, Akashi Sales, and Kawasaki USA Sales in Los Angeles

Production plan
- The basis of the plant budget
- Eighteen-month horizon for capacity projection purposes. The last three to six months are based on an in-plant forecast
- Three-month replanning cycle
- Three months are frozen by the agreement on the sales plan
- Six weeks are absolutely fixed by the shipping lead time from Akashi

Daily production schedule
- This is a final assembly schedule by days. It serves as the Master Production Schedule
- The planning horizon is one year, but only the earliest 4-5 months are given planning attention in detail
- A major replanning cycle occurs every three months corresponding to the planning cycle of the Sales Plan and Production Plan
- A minor replanning cycle occurs weekly
- Three months are frozen by the Sales Plan and also by the planning horizon at Akashi from which parts are shipped
- Six to seven weeks are frozen unchangeably because of agreement with the Akashi Parts Kitting Warehouse on the sequence of kits to be shipped

Fabrication schedules
- These are developed by manually exploding and offsetting the schedules from the Daily Production Schedule (Final Assembly Schedule)
- Because fabrication rates are not yet balanced with assembly rates, it is necessary to build frame and gas tank buffer stocks ahead of final assembly. This results in a build up and draw down of inventory
- One or two day's inventory ahead of final assembly is usually necessary
- Parts Kits from Akashi are ordered on the basis of the Daily Production Schedule. The Kit explosion is done in Akashi
- Parts from other suppliers are ordered on the basis of a manual explosion of the Daily Production Schedule and the Fabrication Schedules
- Each production planner is responsible for the manual explosions of the model schedules for which he/she is responsible and for seeing that materials are on order
- Major replanning cycle each three months. Minor replanning cycle weekly

Planning data
- There are three bills of material:
  (1) Engineering bill of material
  (2) Manufacturing bill of material – used for backflush of WIP based on the count of complete units transferred to the Distribution Warehouse
  (3) Assembly or 'pick list' bill – this is the list of parts needed for final assembly
- Capacity planning is done manually on the basis of historical standards. They keep a record of hours by part number in fabrication and of assembly hours by model
- Records of buffer inventories on the floor are based on production counts at the following points:
  (1) After frame welding
  (2) After fuel tank painting
  (3) After assembly (transfers to Distribution)
- Engineering changes are coded by the production planner responsible. On the engineering bill the code is as follows:
  *Part being phased out
  **Part being phased in
- Most changes are timed to occur with the beginning of a specific lot and the following data is coded on the manufacturing bill:
  (1) Effective date
  (2) Run out quantities of the phased out part
  (3) Serial number of the unit on which the change will be effective (Just starting)

Fig. 4 Planning system of the Lincoln Plant (as of May 1981)

Unlike other products in the Lincoln Plant, the KLT was designed in Los Angeles, not in Japan. The first units were not built in Akashi, but at Lincoln, the first new model to start up in that plant.

Japanese methods of design play a role in how new models start up in Japanese plants, including the Akashi Works. Prints specify materials and dimensions critical to safety performance and reliability. Selection of exact fastener types or non-critical parts may be left to negotiations with suppliers. Details of non-critical parts fit may not be determined until plant production starts. For example, some Kawasaki parts called only for 'an appropriate polymer coating'.

This practice is somewhat disconcerting to American suppliers. Purchasing agents ask for supplier recommendations on materials and minor configuration changes. After recovering from their surprise, most suppliers make recommendations and negotiate a contract, but some refuse to deliver until they have a detailed print to cover them for product liability. A few refuse to negotiate at all.

A new model start up in Japan takes place with design engineers living in the plant and working out the changes. New models in Akashi typically go through one or two 'pilot production runs' to work out changes. (Japanese auto companies may make several thousand minor design changes during start up of a completely redesigned model.)

The start up of the KLT in Lincoln followed an Americanised version of the Japanese pattern. The total time from the initial design concept to running production was seven months. There were 27 running changes during start up. Compared to American practice, the final design was consolidated during start up.

The KLT line assembles only the KLT Model, of which there are two engine size variations. It is shorter than the main line and moves at a slower rate, but because the KLT was born in the Lincoln plant, the KLT line has incorporated many 'Nebraska-grown' ideas into Japanese methods of manufacturing.

Production planning system

Fig. 4 outlines the production planning system of the Lincoln plant as it existed in May 1981. According to the production planning manager changes in the system were occurring so rapidly that they did not plan by exactly the same method for more than three months at a time.

The planning system contains several elements of the MRP system which had been in development. The final assembly schedule serves as the master production schedule (Fig. 5). From this, other plans are derived by exploding the schedules manually. An example of this is the frame welding schedule in Fig. 6. Note that the daily rate of frame welding is well below the rate of frame use in assembly, so the build up of buffer inventory must be time offset from the requirements of the assembly schedule.

Production week	DAY	#5 6/1	#1 6/8	#2 6/15	#3 6/22	#4 6/29/96	Total	Date 3·26·81
Model				(3 WHEEL BIKE)				
T250	Order No							
	QTY	60 60 60 60 60	60 60 60 60 60	60 60 60 60	60 60 60 60	60 60	1320	
T200	Order No							
	QTY							
I250	Order No							
	QTY							
400	Order No							
	QTY					(MAIN LINE SCHEDULE)			
M400	Order No	40	220 40	120 80	100 100		1000	
	QTY	40						
500	Order No	60 20 20 20 20		40 60	280 20		1440	
	QTY			60				
J650	Order No		50 40	40 20 40	100	100	800	
	QTY							
650A	Order No		220 180	60	100 110		710	
	QTY							
700	Order No				50 100 50		200	
	QTY							
00	Order No							
	QTY							
M400	Order No				(SNOWMOBILE BACK LINE)			
	QTY							
M600	Order No	15 25 35 35	35 35 35 35 35	35 35 35 35 35	35 35 35 35	30 30	695	
	QTY							
00	Order No							
	QTY							

Fig. 5 Daily production schedule

		7-20	21	22	23	24
Planned build						
Balance on hand		50	220	130		
Release		375	155	25	25	25
Lot No./balance parts						200
Order No.		SSA W07 →				

		7-27	28	29	30	31
Planned build						
Balance on hand		83	141	149	215	215
Release		58	58	58	26	
Lot No./balance parts		142	58	26		
Order No.		SSA W07 →				

		8-3	4	5	6	7
Planned build						
Balance on hand		160	40			
Release		65	25	25	25	25
Lot No./balance parts						
Order No.		SSA W08 →				

		8-10	11	12	13	14
Planned build						
Balance on hand						
Release		25	25	25	25	25
Lot No./balance parts						
Order No.		SSA W08 →				

		8-17	18	19	20	21
Planned build						
Balance on hand		51	109	167	110	25
Release		26	58	58	58	7
Lot No./balance parts					115	85
Order No.		SSA W08 →				

		8-24	25	26	27	28
Planned build						
Balance on hand		25	25	25	25	25
Release						
Lot No./balance parts						
Order No.						

Fig. 6 Frame welding schedule

A manual explosion also develops purchasing requirements. Each production planner develops purchase release schedules from the final assembly schedules and the fabrication schedules. These schedules go to purchasing to be built into the blanket purchase agreement or to use as revisions to previous delivery schedules.

It was unnecessary to explode requirements at Lincoln for the parts shipped from Akashi. The final assembly schedule was sent to Akashi and the requirements to be shipped were exploded in Japan. Therefore the planning bill of materials for most of the materials planning in Lincoln included only non-Akashi parts.

Since the volume of the Lincoln plant was small compared with the volume of the Akashi Works, Akashi could treat parts orders from Lincoln almost as parts orders from any other customer. Lincoln provided Akashi with schedules in advance, and Akashi could ship the parts as long as deviation from the schedule was not substantial. The real shipping requirements were established by cabling the assembly schedule for Lincoln to Akashi six weeks ahead of time. No special link had been developed between the Lincoln master schedule and the Akashi master schedule.

Sales of all products were seasonal. By building inventory ahead of the selling seasons, Kawasaki levelled the production schedules of the Lincoln plant. All finished units transferred immediately to distribution, and the plant had no further responsibility for them. Without detailed records, plant personnel estimated finished motorcycle inventories at 18–20 days in May 1981.

The Lincoln final assembly schedule (master schedule) was nearly level each day in units made and in total workload, but it was not an identical schedule every day. Only the KLT assembly schedule was identical every day, and that was because KLT 250 was the only model running most of the year. The main motorcycle assembly line schedule averaged 200 motorcycles a day, and the lot sizes were set at 200. However, the number of motorcycles scheduled each day varied from 200 depending on whether a model change was scheduled and on how difficult that change would be.

The lot size of 200 was common to all motorcycles, whether assembled on the main line or not. Lot sizes for the KLT were 200, and that line was scheduled for 60 per day.

The lot size for many snowmobile and jet ski parts was also 200, but not for all of them. The planning for uniform lot sizes on those products had just begun, and the schedule planning was complicated because weather was a great factor in the sales of snowmobiles. In November 1980, the season's schedule of snowmobile production was drastically reduced because snowfall over North America was forecasted to be well below normal for the coming winter. In June it was necessary to rework many unsold snowmobiles in stock to refit them as 1982 models. The prospects of a level schedule assembling snowmobiles seemed difficult to ensure. Motorcycle demand was more predictable so the early efforts to reduce set-up times and balance operation rates concentrated on motorcycles.

Sometimes the schedule was adjusted as the Lincoln plant moved ahead of the plan or behind it. According to foremen, if the daily assembly schedule was met early in the day, the lines continued to work until the close of the shift. If unfavourable events put them behind schedule, they would try to catch up as quickly as possible, sometimes by working overtime. However, in early 1981 sales were so sluggish that the usual problem was working ahead of schedule. If everything went smoothly, the lines got several days ahead of schedule. They stopped assembly operations only when they ran out of material. At that point, people had a choice of going home or staying to do clean up work.

In April 1981 the plant ran ahead of schedule deliberately in order to test how much they could reduce the 'float' of material shipped from Akashi. They found that they needed a two or three day float on hand to allow for the variance in shipping times from Japan.

Getting the help of the suppliers

At Lincoln, labour cost is a small percentage of the total cost of a finished motorcycle. The major pay-off from improved productivity is from improved quality, reduced scrap and reduced inventory investment. To get the maximum reduction in inventory, Kawasaki needed the cooperation of their suppliers in reducing the quantities and increasing the frequency of shipments. Improving the quality of purchased materials while reducing the inventory became a major challenge for the Lincoln purchasing department.

A supplier which needed no instruction was Tricon, a subsidiary of Tokyo Seat Company, located two miles from the Lincoln plant. The three top managers of Tricon were Japanese, and they found it easy to operate the Japanese way with the Lincoln plant. Kawasaki gave them the Lincoln assembly schedules in advance and then telephoned every morning with any changes. Twice daily Tricon delivered the exact number of seats requested to the receiving department, not directly to the assembly lines as in Japan.

Tricon itself bought materials from Japan. They purchased in small quantities from American suppliers and operated with very short set-up times. When the Lincoln plant got off schedule with delays followed by catch up, Tricon had to operate the same way. In turn, they passed the problem to their own suppliers, both Japanese and American.

Until 1980, Kawasaki-Lincoln typically issued blanket purchase orders covering three months and specifying shipping release dates and quantities. Their suppliers understood and accepted these terms. Many American purchase contracts are similiar.

By May 1981 eight suppliers, including Tricon, had agreed to what were called Just-in-time purchasing agreements. These were three-month blanket orders as before, but with three major changes:

- Shipments were to arrive at least twice a week.
- The purchase releases specified *arrival* dates for specified quantities to arrive at the Lincoln plant, not shipping dates as before.
- The accepted deviation from shipping quantity was ±1% of the quantity requested, not ±10% as before.

To be classified as a just-in-time supplier, all three criteria had to be met, but the purchasing agents tried to get as many suppliers as possible to agree to one or more of the conditions.

All three conditions required missionary development of the suppliers. The first reaction to the request for smaller shipment sizes was usually that it was possible, but would cost more. Suppliers presumed that Kawasaki wanted others to carry their inventory for them. Getting suppliers to understand that Kawasaki would like them to cut their own set-up times and costs was a very difficult sales task.

Suppliers also thought it strange that Kawasaki wanted them to be responsible for delivery on a particular date. Americans usually assume that the purchasing plant has taken title as soon as material leaves the shipping dock of their suppliers, and the purchaser shares responsibility for tracking a shipment to see that it arrives as planned. Kawasaki became an exception to its suppliers' standard rules. Most of them could comply, but needed to be reminded occasionally of Kawasaki's custom.

Most suppliers also could not understand why Kawasaki demanded an exact number of parts in each shipment which was only one of a long series of shipments. Getting them to package in standard lot sizes of 200 ready to take to the lines took even more explanation.

Kawasaki first had to educate its own buyers, each of whom had to be unshakably familiar with the reasons for Kawasaki's practices. Representatives of the suppliers daily added to the list of new ways to misunderstand. In order to obtain better working relationships with the suppliers, Kawasaki reduced the number. The idea was to develop much better feedback on quality and scheduling problems with a limited number of suppliers rather than to retain a large number of suppliers each of whom could be threatened with a cut-off in case of problems. For example, Kawasaki cut back the number of its suppliers for polymeric parts from 12 to four in order to obtain a small number of efficient and responsive suppliers rather than a large number of inefficient ones.

Selection of suppliers to retain depended on several criteria:

- Willingness to work with Kawasaki. Buyers explained the JIT purchasing criteria to suppliers who either responded that they were interested in learning more about it or they were not. If they were not interested, discussion ceased. If they were interested, an educational programme began.

- Size of the supplier. A supplier selling Kawasaki 50% of its output is more responsive than a supplier selling Kawasaki less than 1% of total output.

- Location of the supplier. Kawasaki prefers suppliers to be nearby so that quality assurance visits are easy to arrange. A nearby location also makes it more feasible to run a regular truck route to pick up from several different suppliers in one day, thus keeping delivery quantities small.

In addition to trying to persuade some of the suppliers to accept all of the conditions of Just-in-time purchasing, Kawasaki attempted to get *all* the suppliers to respond to arrival dates on the shipping releases and ship within ±1% of the release quantity. Although most suppliers could observe the arrival dates, many could not or would not meet the request for exact shipping quantities.

An example of the type of relationship Kawasaki wanted to build is the one with Brownie Manufacturing Company only a few miles from Lincoln. Kawasaki has an agreement with Brownie to deliver a few thousand pounds of steel tubing for motorcycle frames once a week. The purchase quantity will be higher after Kawasaki works off its tubing inventory from the days when the company felt it had to protect itself from unreliable delivery promises.

By dealing exclusively with Brownie, Kawasaki hopes both to improve the quality of tubing and to reduce tubing inventory. Non-uniform tubing causes problems in frame welding, and the production of tubing is subject to many variables beginning with the quality of the coil stock from which it is made. By working closely with Brownie, the Kawasaki quality control department helps Brownie establish the causes of problems that arise in welding frames. By keeping the delivery quantities small, Kawasaki also provides Brownie much quicker feedback on problems, and can properly identify which lots of tubing have presented problems.

Brownie's management had trouble understanding why Kawasaki wanted exact delivery quantities until they realised that by matching tubing deliveries with Kawasaki lot processing requirements, tracing the source of quality problems is less difficult.

Kawasaki expects to buy about 30% of Brownie's total production, so Brownie listens. At the outset, Brownie was not the supplier of the highest quality tubing, but by working with this company over time Kawasaki hopes to achieve exactly what it wants.

It was extremely hard for people in the supplier companies to understand the logic behind Kawasaki's requests. Several suppliers were invited to the Lincoln plant to observe operations for themselves.

The purchasing department planned to hold 'suppliers' days' from time to time in order to bring together a number of suppliers at once and show them why Kawasaki wanted Just-in-time purchasing. Kawasaki hoped also to show them how to save money for themselves and for Kawasaki by reducing set-up times themselves. Demonstrations are the most effective form of persuasion.

The companies which had decreased shipping quantities did not increase prices much, but results were hard to estimate because everyone increased their prices during times of inflation. Material costs from the Just-in-time purchasing agreements were estimated to have increased no more than 1% above what they would have been otherwise.

Keeping control of shipping costs while decreasing shipping quantities was very important in attaining real benefits. Kawasaki took pains to have the same truck pick up from more than one supplier if that would help. It was in effect beginning to organise some 'truck routes' for regular pick up.

To decrease the cost of freight from a group of suppliers in Minneapolis, Kawasaki had the suppliers deliver in small quantities to a consolidation warehouse where the deliveries were combined into one truckload for the 300 mile trip to Lincoln.

The human factors which make it work

One of the key moves was expanding the responsibility which everyone at the Lincoln plant exercised on their job. Although not completely developed, the Lincoln plant adopted the policy that people must work together and also be responsible for the results of their individual work. Management was responsible for providing the means to overcome obstacles to doing work properly.

What do the people on the plant floor want? The question was asked several times, and the response was always some variation of the following:

- Job security; no layoffs as soon as things get the least bit slack.
- Reasonable pay.
- Attention from management; someone who will listen.

The attention from management is what makes Kawasaki different from many American factories. It begins with the foremen, and conversation with them is revealing:

On personnel development: "All my people are different, and I have to bring them along differently. Some want a bigger job, or they want to try different jobs. Other just want to be left alone and stick with the same thing. I try to change around the ones who like that."

On worker discipline: "Peer pressure is a good thing, and we have a little of that. If somebody screws up and other people have to bust themselves to cover for it, they let them know. It's better than me jumping them."

On treating the workers as responsible people and listening to them: "Everybody knows if you really mean it or are just pretending."

Getting staff people who will work with floor personnel and listen to them is not always easy, but it is a crucial step in getting the floor workforce to realise that they are paid to work with their heads and not just with their hands. Kawasaki had made some headway on that. Engineers were selected for their willingness to listen to suggestions and not play the role of unapproachable expert. This style of behaviour required practice from those unaccustomed to it.

Development of people seems to be the management theme of the Lincoln plant. Dennis Butt stressed that he is trying to develop his managers. The managers and engineers in turn say that their job 'is to make the lights turn on among the people on the shop-floor', and the foremen want to develop their people.

The style was very informal; so informal that Dennis Butt feared quality control circles would ruin the atmosphere with formality. "They aren't going to go for any forced meeting stuff. Our people will respond as individuals if we just point 'em in the direction." Plant improvements will go at the rate which worker development allows and take its own course.

The role of the Japanese advisors in the Lincoln plant was an indirect one. Officially, they did the communicating with the Akashi Works, and they interpreted Japanese prints and documents. Unofficially, their suggestions for plant improvement carried weight. The senior Japanese advisor had 17 years' experience in manufacturing engineering and he had participated in the early development of Kanban at Akashi. His suggestions on equipment changes always received serious attention, but on matters of dealing with the workforce, the Japanese were always at a disadvantage although over time they came to have a much better understanding of Americans. Their influence was important for setting direction. They knew it was vital to have workforce participation, to have flexible workers and to develop the people to work precisely.

Future direction of the Lincoln plant

Despite great progress, the Lincoln plant had fallen short of its goals. Kawasaki had wanted to have some kanban cards in circulation by June 1981, but the fabrication rates were not yet balanced enough with assembly rates to do it.

Kawasaki had an ongoing goal of continuing to work down its inventory to reach 26 inventory turns per year in 1982. That would be possible only if many more suppliers were converted to Just-in-time agreements.

The company wanted to use the space released by the reduction of inventory to continue plant expansion and to put as many operations as possible directly on the final assembly lines. Failing that, it wanted to cut the buffer inventory between operations and final assembly to zero, or very near zero.

By Fall 1981 Dennis Butt wanted to be alternating at least two models in final assembly on the main motorcycle line. This would be a step toward the practice in Akashi of switching models in final assembly after every five units.

Dennis Butt also expressed great confidence in the Lincoln workforce. The Lincoln Plant had started from a level of performance well behind that of the Akashi Works. He estimated that for the Lincoln plant to match Akashi in total performance would take five more years. There was no thought of returning to the practices used before Just-in-time.

Commentary

Most of the Japanese-owned manufacturing plants in the USA had previously developed quality improvement programmes with employee involvement in problem solving. They also usually had some form of no-layoff policy. However, none of them had dared to try Japanese methods of productivity improvement and material flow control. Kanban is not universally practiced in Japan, and in 1981 it was just taking strong root among those companies adopting it, with the exception of Toyota family companies. There are a number of questions that can be posed.

Would MRP work at Lincoln?

It probably would have worked had the Lincoln Plant achieved the record accuracy and discipline to make it work in 1979. The Lincoln operations are simple compared with many MRP applications, and most consultants would not regard Lincoln as a very difficult plant in which to install MRP – except for one thing. In 1979 materials were out of control.

The major question at that time was how to get the plant under control in the shortest possible time while boosting the morale of the workforce. Since everyone was rapidly getting nowhere with MRP, they decided to try something which promised improvement much quicker. The major initial goal was to get the right part to the right place at the right time. That they needed to do quickly before the confidence of the workforce was lost forever. The ideas came from Japanese practice. Several of the actions come from American materials management practice.

What is the logic behind the small lot sizes?

If set-up times and costs are nearly zero, there is no justification for large lot sizes. In addition, if there is enough capacity to make whatever is wanted each day in small quantities, there is no reason to build large inventories of parts, subassemblies or finished goods. The inventory is kept at the lowest level raw material, and one makes what is wanted when it is wanted. That is the logic.

The EOQ equation is only the best known of many methods of lot-sizing applications. However, the philosophy behind Kanban is never to accept the cost parameters which lead one to devise an 'optimal' lot size. A superior approach is to work constantly to reduce the costs. Acceptance of the parameters that lead to optimums is acceptance of costs and inefficiencies.

What was the reason for the fixed lot sizes and the designated floor locations?

Fixing the lot sizes and detailing floor locations for parts are the beginning steps in simplifying things to make the Kanban approach begin working. Fully developed, the shop-floor has material locations almost as well defined as a stockroom. Taking this action was the key reason the Lincoln Plant began getting the right part to the right place. The same has been true in Japan.

However, at the time of writing the case, the Lincoln Plant was struck at the 200 level on lot sizes. They had already achieved most of the benefits which that action would force on them and needed to cut lot sizes again for another round of improvement.

How should the Lincoln Plant proceed from this point?

This is the big question because of unknown reactions of Americans. However, in Japan the actions for developing Kanban are primarily in two categories:

- *Eliminate inventory.* Where inventory exists, question why it exists. To make real improvement never live with a comfortable level of inventory on the shop-floor. Everyone should constantly look for ways to live with a lower level of work-in-process inventory. The Lincoln Plant has not really squeezed very much on the shop-floor yet. They need to cut the parts banks and see what will happen.
  (A Japanese manager would not be happy with the results of cutting the float of parts coming from Akashi in the Spring of 1981. He would want to know what can be done to create such a dependable delivery system that the float can be reduced. Live with one day's float until you determine how to live with it.) The same is true inside the plant. For example, they should eliminate the safety stock being held in frames and see what happens.

- *Balance the production rates in all areas.* In Kanban plants people are constantly looking for imbalances and the reason for them. Is an operator hurrying too much to keep up? If so, what is the reason? Work done too fast may lead to scrap and interruption. Work done too slow leads to building ahead and holding inventory. Everyone should concentrate on achieving balanced production by small modifications of existing equipment before buying new equipment.

These very general principles do not tell anyone what they should do tomorrow morning. A necessary condition for taking action is that the workforce understand and accept what is about to happen to them and why. Therefore each step must be accompanied with advance communication about why it is being done. The objective of mixing models on the main line had already been set. The next step would be to reduce lot sizes on that line again and have a campaign to make quick changeovers every time. This would be a campaign to further specify the locations of parts along the assembly line and modify assembly tools to require no changeover time between models. Naturally, if the people on the line were not given time to adjust their emotions in preparation for the step, the programme might backfire.

One of the most important steps, therefore, is a programme to much more completely explain to the staff and floor personnel what the Lincoln Plant is going through. It probably cannot be done without doing as the Japanese do – making problems and objectives highly visible, and announcing in advance what each new step will be. Then the operators can contribute to the solving of the detailed problems.

Why do the industrial engineers not set standards? Are they not needed to measure costs and decide what to do?

If the Kanban procedure is followed, the entire idea is to make problems so obvious that very little data is needed to identify what the problems are. The workforce can see what they are. Another 'principle' is to keep the implementation moving so that people get no opportunity to become so accustomed to problems that they no longer really see them. Also, as inferred by the case itself, if the industrial engineers are kept busy making improvements, they have no time setting the exact standards which are going to be obsolete very shortly after they are set. Cut obvious waste and cost is also cut. Count the cost of making a change only when necessary.

How does Just-in-time production differ from Kanban, or does it?

The Lincoln Plant started an improvement programme based on methods very similar to Japanese Kanban. What they were doing in 1981 is more like MRP with fixed lot sizes than fully developed Kanban, however.

Only on the KLT line were they running a schedule which was almost identical every day. The main line schedules were definitely not level, and they had only made a start levelling the schedule on the back line which mostly built snowmobiles.

A very good issue on the back line is whether a level schedule can realistically be attained. The products are highly seasonal and subject to extreme changes in schedule as occurred in November 1981. However, Fig. 5 shows a close to level schedule.

The schedule on the main line needs to be levelled before they can clearly see other problems feeding that line. Once the schedule is levelled, they need to live by it. Continuing to run after the day's schedule is complete does not attain that goal. (In Japan, when this is done, the workers clean up, prepare for tomorrow's schedule and use any remaining time to review how to improve operations.) The American managers did not understand how to physically organise the assembly lines for mixed model assembly. Parts positioning was still too lax, and containers of some parts were too large. That was done soon after the case was written.

If the schedule is too great and remains incomplete at the end of the day it is doubtful if everyone on the line could remain over time to finish to schedule. Therefore, it should be possible to finish early on most days with extra time spent on the improvement activity.

The methods of making improvement in Lincoln show the handiwork of materials people. The pick lists are clearly an American approach which Japanese would not use. However, they worked and served a useful purpose early in Lincoln's programme. If a stable schedule and small lot sizes cannot be attained on the back line perhaps pick lists will be necessary for a long time on that line. Otherwise, the Lincoln Plant should work toward phasing them out by improving location control on-line and off-line, and also by a training programme for those stocking the lines.

Passing the Foreign Trade Zone audit was a tremendous achievement at Lincoln. All the material control personnel went through a crash programme to upgrade the inventory system to do it. However, the effort required distracted them from full pursuit of Just-in-time production (or Kanban) because other actions were delayed.

The decrease in inventory level in the Lincoln Plant is limited by the variance in shipping times from Akashi. The programme to stabilise material arrival times is obviously subject to much more difficulty than such a programme in Japan.

The fabrication areas decreased lot sizes, but they were really running by an MRP system in 1981. They had to do more work to get frame welding rates equal to assembly rates on the main line. This required additional work to quickly change fixtures for welding frames, or to obtain the same effect in some other way.

The problems in gas tank fabrication were technically the most difficult. They had never been solved at Akashi either. Therefore, Akashi did the next best to Kanban. They set up a very simple MRP schedule close to the action. Lincoln will probably do something similar.

The Lincoln Plant came closest to the ideal of straight-through flow on the KLT line. Because of the infrequency of a model change on that line, it was easier to approach a level schedule, and attacks on the problems were going faster. Several of the ideas implemented were American originals. The main line was a bigger challenge.

What is the connection between quality improvement and Kanban?

Improvements in both quality and productivity follow from the same principle – making problems visible to everyone. By drastically reducing the errors in assembly (and coming into assembly) the ideal of crating motorcycles at the end of the KLT line was becoming a serious thought.

The connection between quality and productivity becomes obvious with Kanban, but it is often buried in confusion otherwise. In fact, one of the major objectives of Kanban is to make quality problems visible. A fast work pace is not desirable because it leads to errors and the need for rework. Doing it right the first time is an idea imbedded in the notion of balancing the production rates of all operations.

Another connection, not often seen at first, is that the production of many small lot sizes requires regular checking of work. Regular checking while doing the operation itself is better than statistical inspection because the causes of problems are more likely to be identified.

Also when processes are closely linked together, the equipment must be maintained in order to prevent stoppages. The regular attention also helps quality.

In short, the idea of Kanban is not to increase the workpace, but to make every move the right one. For many companies, the major objective of JIT is to *stimulate quality improvement.*

Does this system work by just pushing inventory from work-in-process into finished goods?

Not really. By producing a little bit of everything every day, the needs of a distribution inventory can be tracked more closely than when final assembly has a sequence of long runs which build up the finished goods inventory in every model built. After all, customer service level is kept high by having a little bit of anything needed, and it is not improved by having a large quantity of anything not being demanded.

Japanese companies questioned on this did not think that they increased finished goods inventory by going through a Kanban programme, and all thought that they were able to track the market closer because of the greatly reduced lead times.

One practice which many companies have used during Kanban is to build a little finished goods ahead each time they started a particularly tough phase of development. That kept the demands of marketing from causing them to abort a programme if they were having trouble getting some new idea debugged.

What do the 'loose' design specifications have to do with Kanban?

It gives production or purchasing more latitude to make changes which reduce the cost of the final product. If the company had to go through too formal a process to make a small change, the 'bureaucracy' would stifle some of the improvements. Therefore, the design engineers must be involved on the production floor. The final specifications are really determined by the criteria of producibility, so the value engineering phase on new products goes faster. This is very important in a system in which constant attention to manufacturing engineering improvements is vital.

The difference between Japanese design practice and American is much more visible to Americans than it is to Japanese. The Japanese do not comprehend the extent to which relations between design and production are formalised in most Western companies.

The planning connection between Akashi and Lincoln seemed to be very loose. Was there no formal planning connection?

It was very loose. Akashi did not have to consider the volume of the Lincoln Plant explicit in its planning during the early years of the plant because Lincoln's volume was so much less than that in Akashi. It was a problem that needed attention.

Kawasaki recognised planning as being one of its management weaknesses. Kanban requires simple planning, but excellent execution. The Americans use more complex planning to anticipate their problems in execution. The Lincoln management is likely to initiate action to improve planning between Lincoln and Akashi.

How can worker involvement be increased at Lincoln?

This is the nearly unanswerable question. Clearly the Lincoln workforce responds to management attention. Americans differ from Japanese, but not that much. Copying culture is something to do with great caution, and the thinking at Lincoln is just to copy Japanese principles, try to stay close to the workforce and do what comes naturally. No tea ceremonies.

One hazard at any American plant, especially one which begins to exist in a 'fishbowl' as the Lincoln Plant was starting to do, is the continuity of the management and distractions from outside people. Soon after, the plant went through a period of restricted outside publicity while concentrating on internal development.

Lincoln management has discussed starting a suggestion programme. Some Japanese companies have formal payout systems and some do not. In the USA, suggestion systems have had varied results, but many of the plans have not worked well because they were overlaid with too many rules or because people felt they deserved a bonus which they did not get. The Lincoln managers wanted to be sure not to put in a plan with a fatal flaw in it, but at the same time they felt that American workers might respond better if they could see more benefit than a pat on the back for contributing to an improvement programme. (Japanese are cautious about making payouts so big that any individual has a chance of becoming a star performer in bonus payments and incur the envy of his fellows. The same issue has ruined American suggestion programmes.)

The role of the foreman cannot be overemphasised. As currently organised, the foremen in the Lincoln Plant are the key people in bringing about change which requires different work habits. The 'people development' theme of the Lincoln management cannot be neglected. The worker matures into more than just 'doing the job'. They continuously improve how the job is done.

Subsequent to the time of writing the case, the productivity at the Lincoln Plant increased by about 20% in six months. The Americans did not understand how to do several things:

- Layout the assembly line for mixed model production.
- Further reduce set-up times in welding and painting and cross-train workers so that fabrication was balanced with assembly. Then Kanban was introduced.
- Check the plant every morning to be sure it was ready to run. (That cut daily downtime by half.)
- Make quality improvements so as to rid themselves of the stockroom for production material, and consequently the need for pick lists.

This case does not record a situation in which Kawasaki could take much pride. Readers should understand that much better was possible.

IBM

J.M. Ward
IBM (UK) Ltd

IBM was founded in 1911 and started an operating agency in the UK in 1918. The IBM World Trade Corporation was set up in 1949, to deal with IBM business outside the USA, closely followed by the founding of IBM UK Ltd in 1951. Today, the IBM Corporation has grown to be a $50 billion revenue business, employing approximately 400,000 people.

The Corporation is split into three major divisions: the US division covering Canada and North America; EMEA covering Europe, Middle East and Africa; and AFE covering South America and the Far East. Products are developed in a single development location, for manufacture in each of the three divisions, with one manufacturing location being given the responsibility of prime process development and agreeing a worldwide sourcing plan.

In 1985 IBM UK's revenue grew by 30% to exceed £3 billion with profits before tax of £520 million. Export of goods and services amounted to £1.6 billion; employees within the UK total 18,798.

In Europe, Havant is one of 15 manufacturing locations, with each plant specialising in certain business areas and technologies. IBM Havant employs 1840 people, and occupies over 1 million ft^2 of office and manufacturing space.

The products in Havant fall into three areas: small and intermediate files, telecommunications, and finance systems. Production consists of predominantly assembly and test operations, with some vertical integration of card assembly/test and flexi-circuits.

The organisation within the plant is continually evolving, with the thrust being towards business units aligned by product area. One of the groups supporting the business units is a Technical and Strategic Programmes group, with senior staff providing focal points for programmes such as continuous flow manufacturing (JIT), logistics and quality improvement programmes.

Finally, the personnel policies that IBM operates include management by objectives, pay and progress based on merit and supported by an appraisal and counselling programme, and employee participation is encouraged through participation in improvement and suggestion programmes.

Quality – A prerequisite

In 1980 IBM Corporation launched a 'quality' drive that was to have far reaching effects on all aspects of the business.

It was in 1980 that quality circles were introduced as the first of many quality improvement techniques, and this helped lay the foundations of an attitude to continuous improvement which has become a way of life.

The Havant Excel Programme commenced in 1981, with a planned approach of encouraging improvement initiatives throughout the plant, through visible recording and recognition of performance. The plant sets approximately 30 Excel targets each year, and to achieve these, performance has to exceed that committed in the operating plan. Every department performs department purpose analysis (DPA) to avoid divergence from the department's assigned responsibilities. It is to a clear purpose statement that every department has aligned its Excel targets.

In 1984, circles activity was extended to a more informal 'improvement project' activity, whereby groups of people were encouraged to work on an idea using a number of techniques, from brainstorming to statistical process control.

Quality performance in all areas has shown continuous improvement since 1981; by year end 1985 vendor conformance was 99.5%, manufacturing service impacts <2% and field quality results improved by 38%.

Throughout 1985 a recognition programme had been running, whereby every quarter each function nominates an improvement activity to be considered for the plant director's award; the winners are judged by the senior management team and recognised at a special luncheon.

It is the plant's involvement of all the workforce, that formed a strong base for the continuous flow manufacturing programme.

Continuous flow manufacturing (CFM)

The IBM Corporation set four goals for the 1980s, one of which was to become the low-cost producer. In order for manufacturing to achieve lower costs, CFM was seen as an essential step.

As a manufacturing plant there was already a positive attitude to continuous improvement through the quality improvement programmes, and good quality and 100% box deliveries were being achieved; thus quality and delivery requirements were being achieved. The challenge of CFM was to maintain this performance, but at lower cost!

In early 1984, an education package was sent to all IBM manufacturing plants to inform them of the Corporation's CFM objectives: total quality control, involvement, and elimination of waste. Supporting these three principles were a series of techniques providing a framework from which to develop local guidelines.

Each plant has interpreted the corporate message in its own particular way, while staying within the general framework/guidelines defined. Havant has developed a set of ten related techniques, the most important of which are those related to elimination of waste.

Through reducing the in-house profile, and gaining more effective asset utilisation the aim was to increase manufacturing flexibility, thus reducing customer lead-time.

The main techniques in achieving CFM were as follows:

- *WIP reduction* – cuts costs and uncovers productivity problems otherwise 'hidden' by costly inventory. The goal is a batch size of one.

- *Group technology* – sees production as one continuous process from receipt to shipment. Ideally it is one location giving flexibility of space and manpower through the process. This eliminates long distances between operations, large WIP inventory and difficult communications created through having specialised departments.

- *Balanced/mixed production* – aims for the cycle time to equal the raw process time. Balanced production needs flexible space, equipment and manpower to match order demand. Total flexibility results from mixed production where resources can be moved from one model or product type to the next according to demand.

- *Kanban* – to manage material movement. It can be anything from an empty bin to a phone call to a vendor or a reminder card. It triggers action when material is pulled through the process from one operation to the next.

- *Tightly coupled logistics* – tightening of logistics from supplier to customer, giving rise to better visibility of requirements, greater awareness of problems, and improved communications.

- *Supplier integration* – good relationships between the plant and outside vendors who are seen as an extension of the whole manufacturing process.

- *Zero defects* – CFM increases the emphasis. People responsible for manufacturing also carry out quality control, eliminating the need for inspection and auditing.

- *Management by sight* – CFM materials flow through layouts designed so that problems become immediately obvious.

- *Multi-skilled people* – CFM at its best means people are masters of several trades, contributing to line balancing. Symptoms are seen first in manufacturing where they should be dealt with as they arise.

- *Focus team* – CFM success depends on cross-functional improvement, from personnel to industrial engineering, finance to manufacturing; teamwork through cooperation.

Communication and education are vital to the success of CFM. Several seminars were held in 1984 for the whole range of levels and disciplines within the plant. The first of three videos was produced in mid-1984 accompanied by a leaflet to explain CFM. This was followed by another video at the end of 1984, focusing on three particular pilot projects which had been implemented in the plant. All 1800 people were invited to view the videos in groups, with opportunities to ask questions at the end of each viewing.

In 1985, awareness and education through the plant has been extended by newsheets, poster campaigns and modules in management development courses. But the emphasis in 1985 has been extending this understanding out to IBM's vendor base.

CFM affects the *total manufacturing process*, from the time orders are placed with vendors to the time the product has been accepted and performs reliably in a customer's office. It is not only those parts of the process primarily concerned with material flow, but design and development together with order management that must be influenced to achieve the best results.

IBM Havant's approach has been focused upon pilot projects within its in-house process on existing product lines. These projects have had to be supported by business cases, to ensure that tangible as well as intangible benefits can be assessed.

Learning by experience

After the initial education and the first video, a number of spontaneous improvement projects started up, focusing on piloting one or more of the techniques on existing products that had *not* been designed with CFM process requirements in mind. To help in the early stages, a focal point was established to act as consultant and catalyst to the various ideas/pilot project teams. The teams were given clear objectives to reduce one or more of the following: WIP, cycle time, set-up time, machine movement. The benefits to be quantified were space, inventory, quality and productivity.

Pilot 1 – 8100/4300 kanban

8100/4300 are medium range processors; the build process includes a kitting operation of both mechanical and logic parts, needed due to the high feature content of the product. A week's production of kits would be available on the Friday prior to the 'start-of-build' week of the product, thus at any one time a week's worth of kits would be in WIP.

In order to obtain inventory relief and space savings, kanban was used to call off the kits from the kit store some 2/3 hours before they were needed, thus substantially reducing space ($2100ft^2$), reducing the box process cycle time by one week, and relieving inventory by £1.8 million.

The kanban used was the kit trolley with a card indicating the next kit required. Currently, an improvement team is working on extending the kanban back to the suppliers of frames, to eliminate buffers which are occupying some 4000ft^2 of warehouse space.

Pilot 2 – 4300 group technology (GT)

The idea for this pilot came from the two videos; the build process was made up of a subassembly and final assembly operation taking 16 hours. The manufacturing and engineering departments decided to combine sub- and final assembly, and to break the assembly into four four-hour operations, thus creating a GT cell.

A series of meetings were held to get the process design right, and a pilot was implemented on the 4361 model. Operators were trained to be flexible enough to perform various operations; line balancing was achieved by teamwork, such that test personnel often helped assemblers complete the machines while waiting for units to test.

Communications were improved, and all the problems were highlighted via a meeting on the line at 3pm each day, attended by manufacturing and support personnel. Space of 1500ft^2 was saved and the box process cycle time was reduced from three to two weeks. There was a four-fold improvement in quality performance.

Pilot 3 – Kanban and GT combined

The 4700 products are a family of displays, keyboards and power/logic units that make up a finance industry workstation.

A group technology layout was implemented with kanban squares between each operation; a small parts store for kitting and engineering support was located on line. Parts are delivered to the building and leave as finished products. There is a warning light system in operation to highlight out of-line situations.

Despite a 20% increase in volume, a space reduction of 20% was achieved, the process cycle time reduced from seven to three weeks and machine movement from 1000ft to 200ft.

Pilot 4 – 3725 continuous flow manufacturing

The 3725/6 products are communication controllers. It is on this product line that the most significant pilot to date has been implemented.

The assembly process consisted of approximately 20 hours of assembly work, consisting of subassembly and final assembly operations, with a substantial amount of buffer between each operation.

A cross-functional focus team was set up to investigate how viable a process redesign would be to group activities together into a U-shaped line. The team developed a business case and in parallel set up a small pilot line to test feasibility. The team was headed and driven by manufacturing personnel who were instrumental in managing the project through to installation.

The pilot line was a success and resulted in the implementation of group technology across the whole product. The final solution consists of eight two-hour operations on two U-shaped lines for 3725, and four four-hour operations on one U-shaped line for 3726. Each stage is linked by a kanban square. Operators rotate around operations working as a team, and are continually tuning the operations.

The cost of implementation was over £150,000, but paid back in less than six months through a saving of 14,000ft^2, inventory relief of £2 million (doubling turnover), between 10–20% reduction in direct time, and greater than 60% improvement in quality.

Assemblers have since designed a simple system for time recording and quality measurement within a U-shaped line, and are constantly coming up with new improvement ideas.

Pilot 5 – Inventory reduction

Greater than 15% of IBM Havant output as a plant is termed 'miscellaneous equipment supplies', which includes extra features required after product shipment, documentation, spares, etc. This obviously contributes to the plant inventory significantly. The inventory is consigned to a kitting vendor to assemble kits. The production control group together with the vendor attempted to reduce the amount of inventory being held throughout the total process.

By examining the process flow, space being consumed, and attacking the inventory profile, the value of inventory was reduced from £2.5 million to less than £500,000, space was reduced by 50% and the inventory profile was reduced by 75%. The improvement team is currently working on obtaining the supply of documentation to the vendor only when it's needed through JIT deliveries.

New product CFM implementation

The 5.25 inch disk storage file used in the IBM PC range of products, was the first product to be introduced into the plant after the CFM principles were understood.

From day one the process design was aimed towards CFM; 26 parts were set up on daily delivery with the balance being weekly. All parts were barcoded on receipt from barcodes supplies by IBM to the vendor with purchase orders. Vendor contracts were set up for three months' orders, three months' raw material intent and a further six months' planning.

The file is made in a clean room environment on a series of U-shaped lines with each operation lasting two minutes. Operators have line stop authority and request further parts that are delivered automatically only *when needed*. Products are shipped from the end of the line direct to the customer.

Improvements

Over a period of 18 months all current products have established pilots running, which have contributed to saving some 27,000ft^2 of manufacturing space, 16,000ft^2 of warehouse space, inventory relief of in excess of £10 million, with substantial increases in productivity and quality performance.

In 1985, there were three key measurements at a plant level which were used to track progress:

- *Process simplification* – a targeted reduction of 25% across all measured products in the inventory profile. This included consignment and stocking through to product shipment. The plant profile was reduced from 31 to 24 days (a 25% reduction).

- *Waste elimination* – it was set out to have 150 improvement projects registered by year end (the final number achieved was 426).

- *Total involvement* – aimed to encourage as many people as possible involved in at least one improvement activity by year end, and achieved a penetration of 86%.

These measurements will continue at a plant level in 1986, with each product group having its own local targets.

Widening the scope – Vendors

With the experience of the in-house pilots, the emphasis in 1985 extended the principles of CFM down the supply lines. This could not have been attempted without the positive quality attitudes of vendors.

The goal of procurement in Havant is to 'develop a vendor base capable of achieving IBM quality and delivery requirements at the lowest ultimate cost'.

Vendor development will be a balance between the IBM requirements (e.g. flexibility and quality of service) and the vendors' needs (e.g. stability and repetitive manufacturing). IBM Havant will develop long-term relationships and a business obligation with vendors, through shared experiences, to cope with the demands of CFM.

Education of not only IBM's vendors, but the company's procurement and purchasing people has been an important investment. All procurement personnel in Havant attended a series of one-day seminars run internally which covered the basics of CFM, external and internal case studies, and workshop activities.

This education was followed up with a series of seven seminars involving 82 vendors by commodity. Videos and case studies were used to explain the theory and practice of CFM and the vendor companies were encouraged to apply the principles to their own pilot project. Follow-up seminars have been held at vendors' premises to take the messages given to senior management down through the organisation to the people on the shop-floor.

Direct line feed of parts will rely on intervention-free receipts with zero quality impacts at the point of use. A golden stock of parts is held, to be made available within 24 hours of problems occurring. Parts packaging is crucial to ensuring ease of count verification, protection throughout process and ease of handling, such that the first time the parts are touched will be on assembly into machines.

In 1986 a substantial increase in the number of parts being delivered direct to the lines will occur; estimates indicate that more than 80% of the value can be covered by JIT delivery of 10% of the parts. This will be supported by an increase in IBM's vendor communications strategy, with the objective of having 20 vendors with direct links by December 1986.

Widening the scope – Development

Supplementing IBM's normal early manufacturing involvement and design for manufacture work, it is important that marketing and distribution are involved as early in the design cycle as possible. The top-level logistics process should be developed and understood before the detailed product design is analysed.

Vendors should be involved prior to engineering design release, with special emphasis on not only the manufacturability and quality, but also the logistics regarding ease of packaging, movement and handling. At year end 1985 there were 11 vendors working with IBM's various development locations.

Parts should be grouped in commodity groups, whereby a vendor can manufacture and collate a kit of parts for a machine, the kit being controlled by one part number. This group kitting coupled with high-level assembly sourcing enables the value of a product to be sourced in few vendors, and maximises the return on JIT deliveries.

Some of the process requirements that help CFM are:

- Standard modular designs – minimum features, few standard models.
- Stable design – eliminate engineering changes post first customer ship.
- Stable ship programmes – machine orders frozen 28 days prior to ship.
- Flexible manufacturing – modular units, cycle time of hours/days.

Future products will be manufactured in generic installations, with maximum flexibility to cope with any changes in process inputs. The products will be developed along the 'family concept', with each product in a set satisfying key family characteristics.

The future

Typical benefits supported by numerous case studies and pilot findings are: inventory reductions of 50–80%, space reductions of more than 50%, productivity increase of 20–50%, and quality improvements of three- to four-fold.

The major gains of JIT have still to come; pilots have been established and have exceeded IBM's expectations regarding tangible benefits. Vendors are learning from IBM as IBM are from vendors through shared practical experiences. A start has been made and it is in 1986 that the results will really begin to come through as IBM's new products lines become established.

High value and bulky parts will be delivered from a small vendor base and received directly onto the manufacturing line in unit pack quantities. Short product cycle times, process consolidation and end-of-line shipment will increase manufacturing flexibility.

Within processes there should be a progressive elimination of the non-value-added activities as material is processed from supplier to customer. Directly involved with this process are those people working within the manufacturing areas.

Continuation of quality programmes coupled with increased emphasis on CFM techniques will help IBM towards its goal to be the low-cost high-quality producer.

Ford Motor Company

C. G. Perkey
Ford Motor Co., USA

Ford Motor Company's Casting Division, like other casting producers, is facing a continuing challenge to improve casting quality and reduce operating costs at greater product/process flexibility.

About three years ago the Casting Division of Ford Motor Company completed an in-depth review of the automotive casting business which included an assessment of the Division's competitiveness, as well as the competitive status of its casting processes and equipment.

The review covered casting operations worldwide. Casting Division management visited 28 iron foundries in nine countries. These foundries produce castings similar to those produced by the Casting Division, utilising various casting processes. Different levels of technology and production systems and utilisation were observed. The success of each foundry operation evaluated seemed to depend equally on the level of technology applied, and management/workforce attitudes.

Casting operations varied from plant to plant, with different locations using slightly different methods and priorities to achieve the same basic goals.

In the USA, the trend seems to be a continuation of large, high-speed, tight-flask moulding-lines, supported by assembly-line-type coremaking and casting finishing.

In Europe there appears to be a tendency toward lower speed production. The introduction of group processing is leading to the reduction of the high-speed assembly line approach. In some respects, Europe seems to lead the USA in the application of new technology: there are 15 impact-moulding lines running in Europe, versus two impact systems in the USA. A large automotive foundry in Europe is also the first to utilise a large vertically parted flaskless-moulding system for the production of cylinder-block castings.

In Japan, a large variety of moulding methods are being utilised, with strong emphasis directed towards quality, reliability and flexibility rather than speed. Like all Japanese plants, the foundries are equipped and laid out for 'Just-in-time' processes. Work efforts concentrate on reducing flash and scrap, and improving yield. The Japanese are also moving towards increased use of vertically parted flaskless-moulding systems; they are presently producing V-6 cylinder blocks on such a system.

As a result of the worldwide foundry operations review, an opportunity plan was developed at Ford Motor Company's Casting Division to maintain state-of-the-art casting processes, improve the company's competitive manufacturing position, achieve best-in-class quality status, and provide an opportunity for further improvements in efficiency and profitability.

Specific plans were developed to establish total process control and also achieve balanced production systems. This required Ford's engineers to take a close look at current systems capacities and bottleneck operations. The application of the JIT concept with its straight-through production and parts process flow provided the necessary information for new production systems and equipment requirements.

High-volume 'Just-in-time' cold-box core production

This section reviews the high-volume Just-in-time core-production process developed for a new 3.0L V6 cylinder-block casting. The Just-in-time process concept (Fig. 1) utilises cold-box cores in a high-volume production system as follows:

- Produce 'consistent quality' cores with improved dimensional stability by using a cold-box binder system. This became necessary to achieve the nominal casting wall-thickness design requirement for the 3.0L V6 cylinder block of 4 ± 0.75mm (0.157 ± 0.030in.).

- Assemble core packages (Fig. 2) in the core room and deliver banded or bolted assemblies to the mould line, resulting in tight dimensional control and minimum variability. This decision was based on Ford's experience with an I4 cylinder-block hot-box core package, which generated wall thickness variability of ± 1.3mm (0.051in.).

- Make necessary changes to existing process flow patterns in the core room to obtain direct flow-through, in-step operation, utilising the Just-in-time concept, within the core-making system and between core-making and moulding. This includes the following:
  –eliminate all in-process core storage.
  –significantly decrease core breakage.
  –reduce core-handling labour.
  –establish direct operator responsibility for continuous supply of high-quality cores to moulding, in quantities required by the moulding operation.

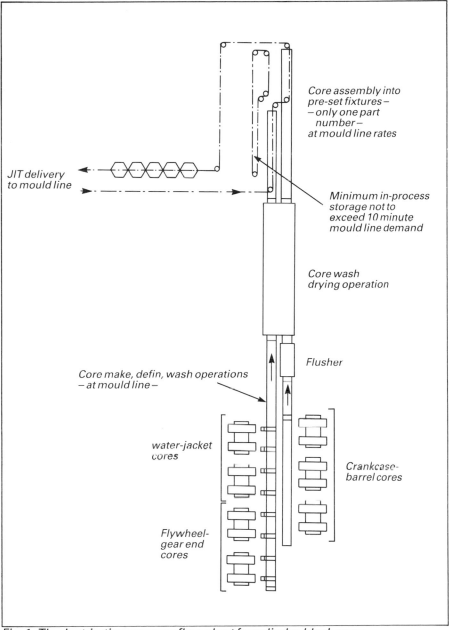

Fig. 1 The Just-in-time process flow-chart for cylinder-block cores

–achieve direct, timely feedback of core-quality information from
    moulding.
–reduce investment for equipment and production systems by not
    providing the presently customary back-up to cover for core
    breakage and machine downtime.

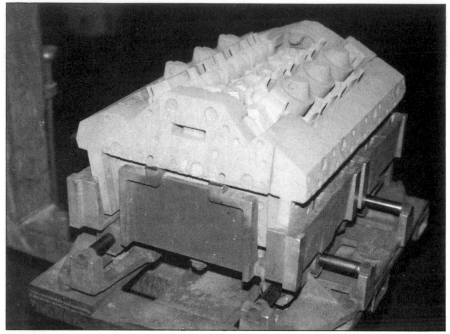

Fig. 2 A core package assembled

- Production system layouts must reflect the flow-through concept; this includes the replacement of monorail hanger-conveyor systems with flow-through belt-conveyor systems, to eliminate manual core-transfer points.
- Core machines must have corebox quick-change systems to assure high-quality core production at maximum machine uptime.
- Systems maintenance must achieve high equipment uptime, since there are no back-up facilities.

An advanced production system for 3.0L V6 cylinder-block cores

At the Cleveland Casting Plant of Ford Motor's Casting Division, an advanced production system for 3.0L V6 cylinder-block cores has been put into operation (Fig. 3). The system was launched in the middle of 1985, for 1986 model-year production.

The new core system was developed to achieve an improved level of core quality at a significant reduction in direct labour. It represents a departure from existing hot-box core-making processes and customary large in-process core inventories.

The following cores are required for one core assembly: three crankcase barrel cores, two side cores, two water jacket cores, one cam tunnel core, one flywheel end core, and one gear end core. The core

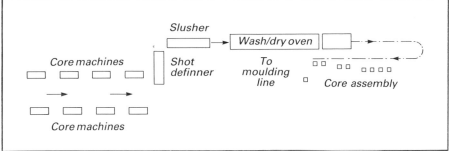

Fig. 3 The JIT core process for cylinder-block cores

assembly has been designed for efficient packaging to achieve maximum dimensional stability as demanded by casting design tolerances. The individual core designs were reviewed for production feasibility. It was determined that the water jacket cores with their 60° V-design (Fig. 4) could not be efficiently blown in hot-box. There was also concern about the strength of the core-prints which lock the water jackets into the end cores.

The decision was made to use a cold-box process for all cores, with the following assumptions.

Core production rate

Total JIT core production of 168 3.0L V6 cylinder-block core assemblies per hour is based on the moulding production rate:

Blow cycle: 35 blows per hour per station
Gassing cycle: 5s; purge cycle: 15s
Core-box size: 762 × 1016mm (30 × 40in.)
Process variability: launch experience: ±1mm (0.039in.)
goal: ±0.75mm (0.030in.)

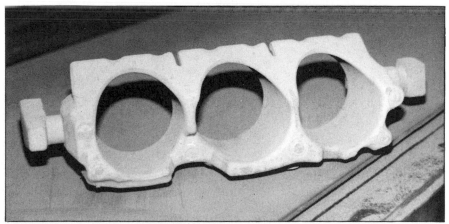

Fig. 4 Water-jacket core for 60° V design

Process flow

- Blow cores – automatic core machine.
- Unload cores from core-blower – manual operation.

Crankcase (side and end cores):
- Transfer cores to belt conveyor – manual operation.
- Remove core fins – automatic definner.
- Apply core-wash – automatic flusher system.
- Dry cores in horizontal wash/dry oven – automatic system.
- Transport cores on belt conveyor to assembly area – automatic.

Water jacket and cam-tunnel cores:

- Remove core fins – manual operation.
- Transfer cores to hanger conveyor – manual operation.
- Apply core-wash – automatic core-wash dip system.
- Dry cores in wash/dry oven – automatic operation.
- Transport cores on hanger conveyor to assembly area – automatic.

All cores:

- Assemble all cores into assembly fixture – manual operation.
- Bolt core assembly – manual operation.
- Transfer core assembly to hanger conveyor shelf – manual operation.
- Deliver core assemblies to mould line via conveyor (+JIT+ mode) – automatic.

Fig. 5 Straight-through flow system for cores

The new core system is a straight-through flow system (Fig. 5) coupled to the mould line. Eight double-station core machines are positioned, four per side, at two common core conveyors. Each core station produces a particular core in support of total core-assembly requirements.

Core machines

The double-station core machines operate on automatic cycle. At the end of a cycle, an automatic pick-up device unloads the cores from the core-box (Fig. 6). The core-machine operator defines the cores individually and loads them onto a common core-conveyor. The operator is responsible for maintaining a steady predetermined production flow of 'OK' cores. To avoid any problems, the operator must monitor the core quality constantly and take immediate action to prevent problem situations. If required, the operator will stop the machine and initiate an automatic core-box quick change. The maximum delay time for such an action is approximately five minutes for a blowplate change plus three minutes for a box change.

Fig. 6 Automatic pick-up device for core unloading

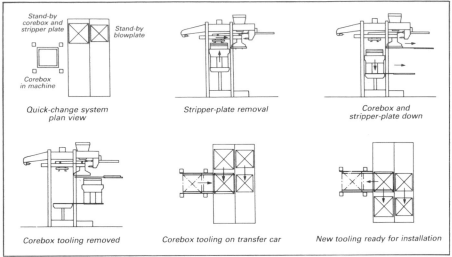

Fig. 7 Corebox quick-change system operating – schematic

Core-box quick-change system operating sequence

A schematic of this system is shown in Fig. 7. The sequence consists of the following operations:

- The replacement core-box and blow-plate are ready on the tooling-shuttle car.
- The tooling-shuttle car (Fig. 8) is moved into position at the core machine for blow-plate removal.

Fig. 8 Corebox quick-change tooling-shuttle car

- The blow-plate removal fixture moves into the core machine under the blow-plate and moves up into position for blow-plate pick-up.
- The blow-plate locking-pins automatically rotate and retract, and release the blow-plate onto the removal fixture.
- The removal fixture transfers the 'used' blow-plate from the operating position under the sand magazine to a holding position on the tooling-shuttle car.
- The 'new' blow-plate is automatically moved into position on the blow-plate transfer fixture.
- The transfer fixture moves into the core machine and up into the position under the sand magazine.
- The locking-pins move and rotate into position, automatically locking the blow-plate into its operating position.
- The 'used' core tooling consisting of upper and lower box and stripper-plate is automatically unlocked by rotating and retracting locking-pins.
- Then the core tooling (upper and lower box and stripper-plate together) moves on power rollers out of its operating position in the core machine into the 'used' tooling position on the tooling-shuttle car.
- The shuttle car then moves to line up the replacement tooling positioned on the other side of the shuttle car with the power rollers on the core machine.
- The replacement tooling is automatically transferred on power rollers (Fig. 9) into the operating position in the core machine.
- Then the locking-pins move and rotate into position, and automatically lock the tooling into place.

Fig. 9 Power rollers for automatic transfer of core tooling

Core-sand system

The sand-preparation plant is designed to mix and deliver cold-box Isocure sand to 16 core-machine stations. The new core-sand system is controlled by three independent programmable controllers (Fig. 10); the two mixers each have their own programmable controller; sand, resin and amine supplies are controlled by an independent third controller.

The central system permits the operator to control all three sections on a fully automatic basis, or by manual push-button control. In the automatic mode, the core-machine stations are being supplied with a pre-set batch weight of sand with either of two adjustable resin contents – on a first-in, first-out basis.

Each core-machine station is being supplied with a pre-selected type of sand upon demand controlled by the probe in the machine hopper (Fig. 11). The probe is set for the minimum amount of sand that permits continuous operation and ensures optimum bench-life by not over-filling a machine hopper.

Sand quality is ensured by the control circuit not permitting any batch mixing cycle to begin unless all conditions of machine function, correct sand weight and resin quantity are satisfied. Quality is further guaranteed by the circuit checking that the sand batch hopper did empty into the mixer and that both resin tanks emptied into the mixer. In the event that not all of the above conditions are being met, the 'bad batch' is automatically discarded, and an alarm is sounded and

Fig. 10 Three independent programmable controllers

Fig. 11 Sand-car receiving core-sand from the machine hopper

acknowledged prior to the operation continuing. Process quality is being ensured by process variables being kept within the following operating tolerances:

- *Raw sand*
  - sand is heated or cooled to 21 ±1.2°C (70 ±4°F) prior to entering the mixer.
  - sand is classified to minimise the build-up of 'fines' 200 mesh and below.
  - batch hopper load-cells control batch size to ±0.5% of set point.
- *Resins – Isocure parts 1 and 2*
  - protected by a nitrogen blanket in storage to minimise the ingress of moisture.
    resins are metered into the resin injection tanks to ±0.5% of set point.
- *Catalyst – amine*
  - protected by a nigrogen blanket in generator day-tanks.
  - catalyst can be proportioned in the range of 5 – 15% by adjustment of carrier-gas weight.
  - generator temperature and pressure controls are adjustable to ±1% accuracy in 2.8°C (5°F) and 3.45kPa (0.5psi) increments respectively.
  - process gas header is heat-traced to maintain 65.5 ±2.2°C (150 ±5°F).
- *Compressed air for blow and purge*
  - dry air, ideally dehumidified to −40°C dewpoint.
  - air-drying capacity is 1132m³/min (4000ft³/min) with an 8.5m³ (300ft³) surge tank.

Operating process detail

The graphic display assists in trouble-shooting system faults, along with a continuous display of system components activity.

The sand-car escort system permits the operator to select which core machine gets the first sand at the beginning of a shift, the last sand near the end of the shift, and to change destination on a given batch of sand as desired during the shift.

The centre console permits the operation of the three gas generators and their heat-traced process-gas header, sand-delivery transporter, the sand heating and cooling system, and the resin delivery system which feeds the resin day-tanks.

Each mixer is set up as follows:

- Batch weight – enter the desired batch weight in pounds for each mixer by using a keypad access code.
- Resin setting – adjust the resin probes to desired resin content for part 1 and part 2 ratio according to developed 'resin tables'. (The 'A' mix code is 1 and the 'B' mix code is 2.)
- Resin code setting – enter the mix type for each machine by the keypad code of A=1 and B=2 for the desired resin content.
- Final mix time setting – enter the final mix time setting in seconds for each mixer by using the keypad access code.
- Bad batch time preset – if any given batch does not receive the proper amount of sand and resins and the sand batch hopper and resin tank did not empty, that batch will automatically be sent to the scrap-sand chute. The bad batch time preset permits the operator to investigate the fault and, if detected and deemed within process tolerance, to cancel the bad batch condition and deliver as programmed. If the bad batch timer 'times out' prior to cancellation, the batch will be automatically rejected.
- All other mixing time elements are programmed and accessible only by the programming panel.
- The sand car can be directed to any core-machine location or to the waste chute.

Present operating parameters

These are as follows:

- All batches = 545kg (1200lb).
- Resin = 1.65 or 1.75%.

Automatic definning

All cores except water-jacket and cam-tunnel cores travel on the common belt conveyor from core-making to and through the automatic definning system (Fig. 12). In the automatic definner the cores are shot-blasted with anti-static plastic beads to remove core fins (Fig. 13).

Fig. 12 Interior view of automatic
definning unit

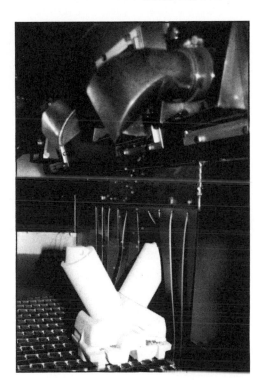

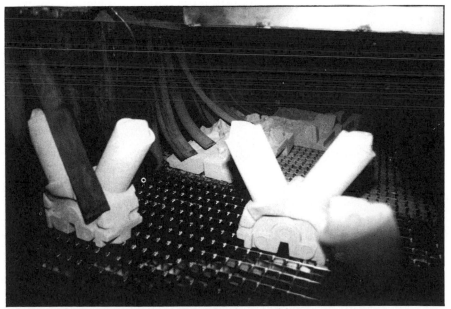

Fig. 13 Definned cores leaving the definning machine

Fig. 14 Core-wash application in the flusher unit

Core-wash application and drying

The next processing station for crankcase, side and end cores is the core-wash flusher (Fig. 14). The cores continue on the common conveyor system through the flusher and then into the core-drying oven (Fig. 15). After the wash-dry operation the cores travel to the assembly area.

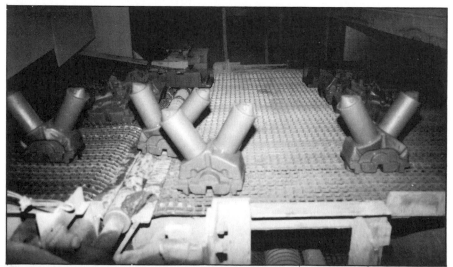

Fig. 15 Automatic transfer of washed cores from flusher into oven

Fig. 16 Custom-designed
hanger conveyor for water
jacket and cam tunnel

Water-jacket cores and cam-tunnel cores, because of their fragile design and lack of a suitable surface that they could be positioned on, are not processed on the common belt conveyor. These cores are handled on a custom-designed hanger conveyor (Fig. 16). This conveyor transports the cores to a core dip station on the mezzanine floor. After core-wash application and drying, the cores are manually transferred to a short run-around hanger conveyor that delivers them to the assembly area.

Core assembly and core delivery

In the assembly area the cores are positioned in preset fixtures which are attached to 'power and free' hangers. The assembly sequence is as follows:

- Set three crankcase-barrel cores.
- Set two end cores.
- Set one cam-tunnel core.
- Close end cores to crankcase and cam-tunnel cores (Fig. 17).

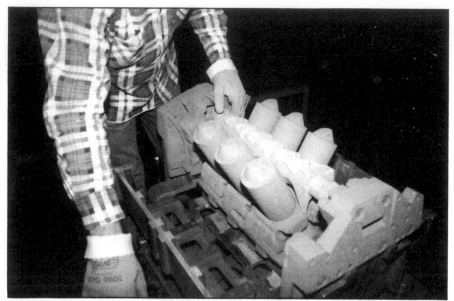

Fig. 17 Partial core assembly

- Set water-jacket cores.
- Set side cores and close to other cores.
- Bolt side cores together with wire bolts in groove in end cores (Fig. 18).

After the assembly operation has been completed, the core package is transferred with a 'transfer assist' device (Fig. 19) onto the foundry conveyor for delivery to the mould line.

Summary of benefits and conclusions

These are preliminary results while the system is still in the launching phase:

- It was proven that JIT flow-through core systems can support mould-line production, and production personnel endorsed the concept.
- Core-machine uptime was improved by 25% with quick-change systems.
- Investment costs were 30% lower, owing to no back-up facilities.
- Core facilities maintenance now receives the same recognition as mould-line maintenance. If the core system goes down, the mould line will go down within 10 minutes.
- A 33% reduction in direct labour was possible owing to elimination of manual core transfers, in-process storage and handling operations.

- Core scrap was reduced by 20%.
- Core-quality problems show up immediately and can be rectified. This avoids a past problem with 'bad' cores hidden in large in-process inventory.

Authors' organisations and addresses

F.J. Bazeley
Ingersoll Engineers
Bourton Hall
Rugby
Warwicks CV23 9SD
England

J.R. Bicheno
School of Mechanical
  Engineering
University of the
  Witwatersrand
1 Jan Smuts Avenue
Johannesburg 2001
Republic of South Africa

J.T. Black
Department of Industrial
  Engineering
Auburn University
308 Dunstan Hall,
AL 36849–3501
USA

W.J. Dumolien
Deere Tech Services
John Deere Road
Moline, IL 61265–8098
USA

R.E. Fox
Avraham Y. Goldratt Institute
57 Trumbull Street
New Haven, CT 06510
USA

D.A. Fulghum
Navistar International Corp.
401 North Michigan Avenue
Chicago, IL 60611
USA

M. Fuse
Plant and Production Engineering
  Department
Sumitomo Electric Industries Ltd
1-3 Shimaya 1-chome
Konohana-ku
Osaka 554
Japan

R.W. Hall
Operations Management
Indiana University
1561 Orchard Road
Mooresville, IN 46158
USA

M.P. Karle
University of the Witwatersrand
1 Jan Smuts Avenue
Johannesburg 2001
Republic of South Africa

D.L. Lee
Cummins Engine Company Ltd
Royal Oak Way South
Daventry
Northants NN1 5NU
England

S.M. Lee
Department of Management
University of Nebraska Lincoln
Lincoln, NE 68588-0491
USA

C. O'Brien
University of Nottingham
Department of Production
  Engineering
Nottingham NG7 2RD
England

C.G. Perkey
Manufacturing and Plant
  Engineering
Casting Division
Ford Motor Company
Dearborn, MI 48121
USA

S. Priestman
RD 2 Box 377
Rhinebeck, NY 12572
USA

R.J. Schonberger
Schonberger & Associates Inc.
PO Box 66948
3235 S.W. 166th Street
Seattle, WA 98166
USA

M. Sepehri
Systems Management Department
University of Southern California
Los Angeles, CA 90089-0021
USA

C.A. Voss
School of Industrial & Business
  Studies
University of Warwick
Coventry CV4 7AL
England

J.M. Ward
IBM United Kingdom Ltd
Havant Plant
PO Box 6
Havant
Hants PO9 1SA
England

V. Winn
Computer Peripherals Bristol
Hewlett-Packard Ltd
Filton Road
Stoke Gifford
Bristol BS12 6QZ
England

J.D. Wood
Lucas Electrical Ltd
Great Hampton Street
Birmingham B18 6AU
England

B. Zimmermann
Maufacturing Engineering
Rank Xerox Manufacturing
  (Nederland) BV
Postbus 43
5800 Venray
The Netherlands

Source of material

Just-in-time: Replacing complexity with simplicity
First published in *Industrial Engineering*, October 1984, pp. 52–63,
under the title 'Just-in-time production systems: Replacing
complexity with simplicity in manufactring management'.
Reprinted courtesy of the author and the American Institute of
Industrial Engineers.

Japanese JIT manufacturing management practices in the UK
First published in *Management World Digest* (The Institution of
Industrial Managers).
Reprinted courtesy of the author and MCB University Press Ltd.

Cellular manufacturing systems
First published in *Industrial Engineering*, November 1983, pp.
36–48, under the title 'Cellular manufacturing systems reduce setup
time, make small lot production economical'.
Reprinted courtesy of the author and the American Institute of
Industrial Engineers.
Updated April 1987.

Cellular manufacturing at John Deere
First published in *Industrial Engineering*, November 1983, pp.
72–75, under the title of 'Cellular manufacturing becomes
philosophy of management at components facility'.
Reprinted courtesy of the authors and the American Institute of
Industrial Engineers.

The Kanban system
First published in *Interfaces*, Vol. 13, No. 4, August 1983, pp. 56–67,
under the title 'Applications of single-card and dual-card Kanban'.
Reprinted courtesy of the author and The Institute of Management
Sciences.

Set-up time reduction: Making JIT work
First published in *Management Services*, May 1986.
Reprinted courtesy of the author and The Institute of Management
Services.
Updated April 1987.

Group technology and JIT
Not previously published.

Developing and implementing control systems for repetitive manufacturing
First published in *Industrial Engineering*, June 1986, pp. 34–36.
Reprinted courtesy of the authors and the American Institute of Industrial Engineers.

OPT: Leapfrogging the Japanese
First published in *Inventories and Production*, March 1983, under the title 'OPT: An answer for America – Part IV, Leapfrogging the Japanese'.
Reprinted courtesy of the author and Wright Publishing Co.
Updated April 1987.

Schedule process management
First published in *Industry Actionline*, November/December 1985, pp. 20–24.
Reprinted courtesy of the author and the Automotive Industry Action Group.
Updated April 1987.

Stockless production: A manufacturing strategy for knowledge society
Not previously published.

SQC and JIT: Partnership in quality
First published in *Quality Progress*, May 1985, pp. 31–34.
Reprinted courtesy of the author and the American Society for Quality Control.

An analysis of Japanese quality control systems
First published in *SAM Advanced Management Journal*, Vol. 18, Part 5, May 1985, pp. 24–31, under the title 'Analysis of Japanese Quality Control Systems: Implications for American manufacturing firms'.
Reprinted courtesy of the authors and the Society for Advancement of Management.

Continuous improvement through standardisation
First published in *Target*, Summer 1986, pp. 3–8.
Reprinted courtesy of the author and the Association for Manufacturing Excellence.

A framework for JIT implementation
Not previously published.

Strategies for implementing JIT

First presented at the 4th European Conference on Automated
Manufacturing, 12–14 May, 1987, Birmingham, UK, pp. 57–73
under the title 'JIT in the corporate strategy'.
Reprinted courtesy of the authors and IFS (Conferences) Ltd.

Comparing Japanese and traditional purchasing

First published in the *International Journal of Operation and
Production Management*, Vol. 5, No. 4, pp. 5–14, under the title
'Comparative analysis of Japanese just-in-time purchasing and
traditional US purchasing systems'.
Reprinted courtesy of the authors and MCB University Press Ltd.

JIT purchasing in the UK

Not previously published.

Lucas Electrical

First published in *Just-in-Time – An Executive Briefing*, pp. 81–87,
1986, under the title 'A systems approach to business redesign'.
Reprinted courtesy of the author.

Rank Xerox

Not previously published (in-house brochure is available from Rank
Xerox).

New United Motor Manufacturing

First published in *Industrial Engineering*, March 1986, pp. 34–41,
under the title 'Car manufacturing joint venture tests feasibility of
Toyota method in US'.
Reprinted courtesy of the author and the American Institute of
Industrial Engineers.

Repco

First published in *Just-in-Time – An Executive Briefing*, pp. 133–140,
1986, under the title 'JIT in the small to medium batch high variety
industry: Repco – A case study'.
Reprinted courtesy of the authors.
Updated April 1987.

An application in the automotive components industry

Not previously published.

Hewlett-Packard

First published in *Industrial Engineering*, February 1986, pp. 44–51, under the title 'Quality and inventory control go hand in hand at Hewlett-Packard's Computer Systems Division'.
Reprinted courtesy of the authors and the American Institute of Industrial Engineers.

Kawasaki USA

First published by the American Production and Inventory Control Society, 1981, under the title 'Kawasaki USA: Transferring Japanese production methods to the USA – A case study'.
Reprinted courtesy of the author and the American Production and Inventory Control Society.
Updated April 1987.

IBM

First published in *Just-In-Time – An Executive Briefing*, pp. 95–99, 1986, under the title 'JIT production: Theory into practice'.
Reprinted courtesy of the author.

Ford Motor Company

First presented at the 1986 BCIRA International Conference on Advances in Technology for Gas-hardened Moulds and Cores, under the title 'Ford Motor: Just-in-time production at the Cleveland Plant'. The bound set of papers are obtainable from BCIRA.
Reprinted courtesy of the author and the British Cast Iron Research Association.